Training Manual
on Transport
and Fluids

Training Manual
on Transport
and Fluids

John C. Neu

Graduate Studies
in Mathematics

Volume 109

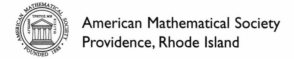

American Mathematical Society
Providence, Rhode Island

2000 *Mathematics Subject Classification.* Primary 35–XX, 44–XX, 76–XX.

For additional information and updates on this book, visit
www.ams.org/bookpages/gsm-109

Library of Congress Cataloging-in-Publication Data
Neu, John C., 1952–
    Training manual on transport and fluids / John C. Neu.
      p. cm. — (Graduate studies in mathematics ; v. 109)
    ISBN 978-0-8218-4083-2 (alk. paper)
    1. Transport theory—Mathematical models. 2. Fluid mechanics—Mathematical models.
I. Title.

QC718.5.T7N48   2998
530.13′8—dc22
                                            2009028296

Contents

Preface

This text presents models of transport in continuous media and a corresponding body of mathematical techniques. Within this text, I have embedded a subtext of problems. Topics and problems are listed together in the table of contents. Each problem is followed by a detailed solution emphasizing process and craftsmanship. These problems and solutions express the *practice* of applied mathematics as the examination and re-examination of essential ideas in many interrelated examples.

Since the science that falls under the headings "transport" or "fluids" is so broad, this introductory text for a one-term advanced undergraduate or beginning graduate course must select a highly specific path. The main requirement is that topics and exercises be logically interconnected and form a self-contained whole.

Briefly, the physical topics are: convection and diffusion as the simplest models of transport, local conservation laws with sources as a general "frame" of continuum mechanics, ideal fluid as the simplest example of an actual physical medium with mass, momentum and energy transport, and finally, free surface waves and shallow water theory. The idea behind this lineup is the progression from purely geometric and kinematic to genuinely physical.

The mathematical prerequisites for engaging the practice of this text are: fluency in advanced calculus and vector analysis, and acquaintance with PDEs from an introductory undergraduate course.

The mathematical skills *developed* in this text have two tracks: First, classical constructions of solutions to linear PDEs and related tools, such as the Dirac δ-function, are presented with a relentless sense of connection to the geometric-physical situations they articulate. Second, and more essential, is the emphasis on dimensional analysis and scaling. Some topics, such as physical similarity and similarity solutions, are traditional. In addition, there are asymptotic reductions based on scaling, such as incompressible flow as a limit of compressible flow, shallow water theory from the full equations of free surface waves, and further reduction of shallow water theory to an asymptotic model of a tsunami approaching the shore. The art of scaling reduction is introduced in the very first chapter, and the drum beat of examples and problems persists throughout the body of the text.

I am grateful to several people who assisted me with the development of this text. Craig Evans asked me to write this book in the first place, and Sergei Gelfand, the editor, reinforced and upheld this request. I thank Tom Beale for discussions and suggestions regarding the title. Katherine Schwarz, a student in my Fall 2004 graduate course on which this book is based, developed the LaTeX version of my lecture notes and figures, many of which are included here. Yu Tanouchi typed the revised and extended version, bringing the book into its present form. Rahel Wachs and Yossi Farjoun made the computerized versions of the figures. Bianca Cerchiai reviewed the book close to its final form and offered many thoughtful suggestions. Finally, I acknowledge my wife, Wanda Krassowska Neu, for her help and support during this work.

The period of writing includes the years from 2005 to 2009, during which I received support from the National Science Foundation grant DMS-0515616.

Part 1

Transport processes: the basic prototypes

Chapter 1

Convection

Material volumes and transport theorem. Let $\rho(\boldsymbol{x}, t)$ and $\boldsymbol{u}(\boldsymbol{x}, t)$ be the density and velocity fields of a fluid. Suppose $\boldsymbol{u}(\boldsymbol{x}, t)$ is given. In principle, trajectories of all fluid particles can be determined, and the density $\rho(\boldsymbol{x}, t)$ at any time t can be determined from an initial condition consisting of given values of $\rho(\boldsymbol{x}, 0)$. To understand this determination, first look at time sequences of regions in space corresponding to the same fluid particles, called *material volumes*. By definition, the total mass of fluid inside a

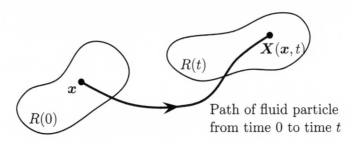

Figure 1.1.

material volume $R(t)$ is independent of t, so

$$(1.1) \qquad \frac{d}{dt} \int_{R(t)} \rho(\boldsymbol{x}, t) \, d\boldsymbol{x} = 0.$$

In general, given a vector field $\boldsymbol{u}(\boldsymbol{x}, t)$ on $\mathbb{R}^n$, one can construct a *flow map* of $\mathbb{R}^n$ into itself. The image of $\boldsymbol{x}$, denoted by $\boldsymbol{X}(\boldsymbol{x}, t)$, satisfies the ODE initial value problem

$$(1.2) \qquad \begin{aligned} \dot{\boldsymbol{X}} &= \boldsymbol{u}(\boldsymbol{x}, t), \quad \text{all } t, \\ \boldsymbol{X}(\boldsymbol{x}, 0) &= \boldsymbol{x}, \quad \text{given.} \end{aligned}$$

The term material volume is henceforth used in a broader sense, referring to the time sequence of images of a fixed region under the flow map.

Let $c(\boldsymbol{x}, t)$ be a real-valued function. We say that $c(\boldsymbol{x}, t)$ is *convected* by the flow $\boldsymbol{u}(\boldsymbol{x}, t)$ if for all material regions $R(t)$ of the flow, we have

$$(1.3) \qquad \frac{d}{dt} \int_{R(t)} c(\boldsymbol{x}, t) \, d\boldsymbol{x} = 0.$$

The heuristic idea is that $c(\boldsymbol{x}, t)$ is the density of some "stuff" which is carried by the flow $\boldsymbol{u}$, such as milk in stirred coffee.

The bulk conservation identity (1.3) implies a PDE for $c(\boldsymbol{x}, t)$. The derivation of the PDE asks for a bit of vector calculus, which we review here. Let $f(\boldsymbol{x}, t)$ be any real-valued function. Then for any material region $R(t)$ of the flow $\boldsymbol{u}(\boldsymbol{x}, t)$,

$$(1.4) \qquad \frac{d}{dt} \int_{R(t)} f(\boldsymbol{x}, t) \, d\boldsymbol{x} = \int_{R(t)} \{f_t + \boldsymbol{\nabla} \cdot (f\boldsymbol{u})\} \, d\boldsymbol{x}.$$

A standard proof of this *transport theorem* uses the flow map to change the variable of space integration in $\int_{R(t)} f(\boldsymbol{x}', t) \, d\boldsymbol{x}'$ from $\boldsymbol{x}'$ ranging over $R(t)$ to $\boldsymbol{x}$ ranging over $R(0)$, and involves time differentiation of the Jacobian determinant. Here, a heuristic calculation better serves a sense of geometric understanding. We have

$$(1.5) \qquad \frac{d}{dt} \int_{R(t)} f \, d\boldsymbol{x} = \int_{R(t)} f_t \, d\boldsymbol{x} + \int_{\partial R} f\boldsymbol{u} \cdot \boldsymbol{n} \, da$$

Here, $\boldsymbol{n}$ is the outward unit normal on $\partial R(t)$. Figure 1.2 is a cartoon to understand the surface integral. In time increment dt, a "material patch" of area da sweeps out a cylinder of volume $(\boldsymbol{u} \cdot \boldsymbol{n} \, dt) \, da = (\boldsymbol{u} \cdot \boldsymbol{n} \, da) \, dt$. If $\boldsymbol{u} \cdot \boldsymbol{n} > 0$, this cylinder is space "annexed" by the material volume. If

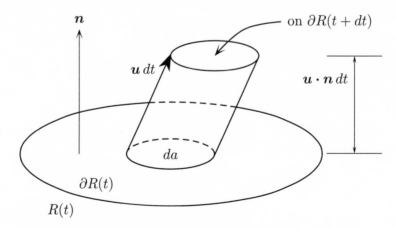

Figure 1.2.

$\boldsymbol{u} \cdot \boldsymbol{n} < 0$, it represents space surrendered. The contribution to $\frac{d}{dt} \int_{R(t)} f \, dx$ due to this cylinder is $f \boldsymbol{u} \cdot \boldsymbol{n} \, da$. Summing over patches da which entirely cover $\partial R(t)$ gives the surface integral in (1.5). The conversion of (1.5) into (1.4) involves an application of the divergence theorem to the surface integral.

A simple corollary of (1.4) identifies the property of the flow $\boldsymbol{u}$ which quantifies local expansion or contraction. In (1.4) take $f \equiv 1$; then we find the rate of change of volume

$$V(t) := \int_{R(t)} 1 \, d\boldsymbol{x}$$

with respect to time to be

$$\dot{V}(t) = \int_{R(t)} \boldsymbol{\nabla} \cdot \boldsymbol{u} \, d\boldsymbol{x}.$$

In the limit of a material volume shrinking to a material point $\boldsymbol{x}(t)$,

$$\frac{1}{V} \int_{R(t)} \boldsymbol{\nabla} \cdot \boldsymbol{u} \, d\boldsymbol{x} \to (\boldsymbol{\nabla} \cdot \boldsymbol{u})(\boldsymbol{x}(t), t)$$

and hence

(1.6) $$\frac{\dot{V}}{V} \to (\boldsymbol{\nabla} \cdot \boldsymbol{u})(\boldsymbol{x}(t), t).$$

Problem 1.1 (Convection along a one-dimensional line). Let x be displacement along the line. The velocity field $u(x, t)$ is a given function of position and time. $(a(t), b(t))$ is a material interval whose endpoints satisfy the ODEs $\dot{a} = u(a, t)$, $\dot{b} = u(b, t)$. Let $f(x, t)$ be any real-valued function of position and time. Formulate the one-dimensional version of the transport theorem (1.4),

$$\frac{d}{dt} \int_{a(t)}^{b(t)} f(x, t) dx = ?$$

and prove it by use of calculus (the fundamental theorem of calculus and rules of differentiation).

Solution. The one-dimensional version of (1.4) is

$$(1.1\text{-}1) \qquad \frac{d}{dt}\int_{a(t)}^{b(t)} f(x,t)dx = \int_{a(t)}^{b(t)} \{f_t + (uf)_x\}dx.$$

To prove (1.1-1), define

$$(1.1\text{-}2) \qquad F(x,t) := \int_0^x f(x',t)dx'.$$

This is an antiderivative of f with respect to x, so we have

$$\int_{a(t)}^{b(t)} f(x,t)dx = F(b(t),t) - F(a(t),t).$$

We differentiate this identity with respect to t, using the chain rule for the right-hand side:

$$(1.1\text{-}3) \qquad \frac{d}{dt}\int_{a(t)}^{b(t)} f(x,t)dx = F_t(b,t) - F_t(a,t) + F_x(b,t)\dot{b} - F_x(a,t)\dot{a}.$$

Time differentiation of (1.1-2) gives

$$F_t(x,t) = \int_0^x f_t(x',t)dx'$$

and hence,

$$(1.1\text{-}4) \qquad F_t(b,t) - F_t(a,t) = \int_0^b f_t(x,t)dx - \int_0^a f_t(x,t)dx = \int_a^b f_t\, dx.$$

Next,

$$(1.1\text{-}5) \qquad F_x(b,t)\dot{b} - F_x(a,t)\dot{a} \;=\; f(b,t)u(b,t) - f(a,t)u(a,t)$$
$$= \int_a^b (uf)_x dx.$$

The first equality uses $F_x = f$, $\dot{a} = u(a,t)$, and $\dot{b} = u(b,t)$. The second equality uses the fundamental theorem of calculus. From (1.1-3)–(1.1-5), it follows that

$$\frac{d}{dt}\int_a^b f\, dx = \int_a^b \{f_t + (uf)_x\}dx.$$

Problem 1.2 (Jacobian of flow map).   Let $\boldsymbol{x} \to \boldsymbol{X}(\boldsymbol{x},t)$ be the flow map of $\mathbb{R}^n$ into itself associated with velocity field $\boldsymbol{u}(\boldsymbol{x},t)$.

a) Show that the flow map's Jacobian determinant $J(\boldsymbol{x},t)$ satisfies

$$(1.2\text{-}1) \qquad J_t(\boldsymbol{x},0) = (\boldsymbol{\nabla} \cdot \boldsymbol{u})(\boldsymbol{x},0).$$

b) Use (1.2-1) to show that the volume $V(t)$ of the material region $R(t)$ satisfies

$$\dot{V}(0) = \int_{R(0)} (\boldsymbol{\nabla} \cdot \boldsymbol{u})(\boldsymbol{x}, 0) d\boldsymbol{x}.$$

This result is invariant under translation of the origin of time:

(1.2-2)
$$\dot{V}(t) = \int_{R(t)} (\boldsymbol{\nabla} \cdot \boldsymbol{u})(\boldsymbol{x}', t) d\boldsymbol{x}'.$$

c) Use (1.2-2) to show that

$$J_t(\boldsymbol{x}, t) = (\boldsymbol{\nabla} \cdot \boldsymbol{u})(\boldsymbol{X}(\boldsymbol{x}, t), t) J(\boldsymbol{x}, t).$$

Solution.

a) We have

$$\boldsymbol{X}(\boldsymbol{x}, h) = \boldsymbol{x} + \boldsymbol{u}(\boldsymbol{x}, 0) h + O(h^2)$$

in the limit of $h \to 0$. In component form,

$$X_i(\boldsymbol{x}, h) = x_i + u_i(\boldsymbol{x}, 0) h + O(h^2).$$

Differentiating with respect to x_j gives the ij component of the Jacobian matrix,

$$\partial_j X_i = \delta_{ij} + (\partial_j u_i)(\boldsymbol{x}, 0) h + O(h^2) = \delta_{ij} + A_{ij} h,$$

where A is the matrix with components

(1.2-3)
$$A_{ij} = (\partial_j u_i)(\boldsymbol{x}, 0) + O(h).$$

Hence, the Jacobian determinant is

(1.2-4)
$$J(\boldsymbol{x}, h) = \mathrm{Det}(I + hA) = 1 + h\mathrm{Tr}A + \cdots + h^n \mathrm{Det}\,A.$$

To prove the second equality, recall the characteristic polynomial of A,

$$p(\lambda) := \mathrm{Det}(\lambda I - A) = \lambda^n - \lambda^{n-1}\mathrm{Tr}A + \cdots + (-1)^n \mathrm{Det}\,A.$$

We compute

$$\mathrm{Det}(I + hA) = (-h)^n p(-\frac{1}{h}) = 1 + h\mathrm{Tr}A + \cdots + h^n \mathrm{Det}\,A.$$

With A as in (1.2-3), (1.2-4) becomes

$$J(\boldsymbol{x}, h) = 1 + h(\partial_{ii} u)(\boldsymbol{x}, 0) + O(h^2),$$

with summation over the repeated index i. In vector notation,

$$J(\boldsymbol{x}, h) = 1 + h(\boldsymbol{\nabla} \cdot \boldsymbol{u})(\boldsymbol{x}, 0) + O(h^2).$$

The coefficient of the 'h' term is $J_t(\boldsymbol{x}, 0)$.

b) We have

(1.2-5) $$V(t) = \int_{R(t)} 1\, d\boldsymbol{x} = \int_{R(0)} J(\boldsymbol{x}, t) d\boldsymbol{x}.$$

Time differentiation and use of (1.2-1) gives

$$\dot{V}(0) = \int_{R(0)} J_t(\boldsymbol{x}, 0) d\boldsymbol{x} = \int_{R(0)} (\boldsymbol{\nabla} \cdot \boldsymbol{u})(\boldsymbol{x}, 0) d\boldsymbol{x}.$$

c) In (1.2-2), change the variable of integration $\boldsymbol{x}'$ in $R(t)$ to the pre-image $\boldsymbol{x}$ in $R(0)$. In this way,

(1.2-6) $$\dot{V}(t) = \int_{R(0)} (\boldsymbol{\nabla} \cdot \boldsymbol{u})(\boldsymbol{X}(\boldsymbol{x}, t), t) J(\boldsymbol{x}, t) d\boldsymbol{x}.$$

However, time differentiation of (1.2-5) gives

(1.2-7) $$\dot{V}(t) = \int_{R(0)} J_t(\boldsymbol{x}, t) d\boldsymbol{x}.$$

Since (1.2-6) and (1.2-7) are both true for arbitrary $R(0)$, we conclude that

$$J_t(\boldsymbol{x}, t) = (\boldsymbol{\nabla} \cdot \boldsymbol{u})(\boldsymbol{X}(\boldsymbol{x}, t), t) J(\boldsymbol{x}, t).$$

Problem 1.3 (Transport theorem by change of variable). Let $R(t)$ be any material region of the flow $\boldsymbol{u}(\boldsymbol{x}, t)$ and let $f(\boldsymbol{x}, t)$ be any real-valued function. Show that

$$\frac{d}{dt} \int_{R(t)} f\, d\boldsymbol{x} = \int_{R(t)} \{f_t + \boldsymbol{\nabla} \cdot (f\boldsymbol{u})\} d\boldsymbol{x}$$

by time differentiation of

$$\int_{R(t)} f(\boldsymbol{x}', t) d\boldsymbol{x}' = \int_{R(0)} f(\boldsymbol{X}(\boldsymbol{x}, t), t) J(\boldsymbol{x}, t) d\boldsymbol{x}.$$

Solution. We have

$$\frac{d}{dt} \int_{R(t)} f(\boldsymbol{x}', t) d\boldsymbol{x}' = \int_{R(0)} \{(f_t + \boldsymbol{X}_t \cdot \boldsymbol{\nabla} f)(\boldsymbol{X}, t) J(\boldsymbol{x}, t) + f(\boldsymbol{X}, t) J_t(\boldsymbol{x}, t)\} d\boldsymbol{x}.$$

In the right-hand side, set $X_t = u(X, t)$ and $J_t(x, t) = (\nabla \cdot u)(X, t)J(x, t)$ from Problem 1.2c, to obtain

$$\frac{d}{dt} \int_{R(t)} f(x', t)dx' = \int_{R(0)} \{(f_t + u \cdot \nabla f + (\nabla \cdot u)f)(X, t)\}J(x, t)dx$$

$$= \int_{R(0)} \{f_t + \nabla \cdot (fu)\}(X, t)J(x, t)dx.$$

In the right-hand side, change the integration variable from x in $R(0)$ to images x' in $R(t)$, to get the final result,

$$\frac{d}{dt} \int_{R(t)} f(x', t)dx' = \int_{R(t)} \{f_t + \nabla \cdot (fu)\}dx'.$$

Dropping the primes gives the transport theorem (1.4).

Problem 1.4 (Time derivative of integral over a changing region). Let $R(t)$ be a time sequence of bounded regions in $\mathbb{R}^n$. The motion of the surface $\partial R(t)$ is quantified by the normal velocity: Think of $x(t)$ as the trajectory of a "bug" trapped on $\partial R(t)$; the projection of its velocity $\dot{x}$ onto the outward unit normal n is the *normal velocity* $v = \dot{x} \cdot n$ of the surface at $x(t)$. For each t, the normal velocity is a function of position on $\partial R(t)$. Show that

$$(1.4\text{-}1) \qquad \frac{d}{dt} \int_{R(t)} f(x, t)dx = \int_{R(t)} f_t\, dx + \int_{\partial R(t)} fv\, da.$$

Solution. We make the (reasonable) assumption that the given time sequence of regions $R(t)$ is a material region of some velocity field $u(x, t)$. Then by Problem 1.3,

$$\frac{d}{dt} \int_R f dx = \int_R \{f_t + \nabla \cdot (fu)\}dx$$

$$= \int_R f_t\, dx + \int_{\partial R} fu \cdot n\, da.$$

The last equality uses the divergence theorem. The restriction of u to ∂R gives the velocities of points on ∂R, so $u \cdot n = v$ and we obtain (1.4-1).

The convection PDE. Now let us return to the consequences of the bulk conservation identity (1.3). From (1.3) and (1.4) it follows that

$$\int_{R(t)} \{c_t + \boldsymbol{\nabla} \cdot (c\boldsymbol{u})\} \, d\boldsymbol{x} = 0$$

for all material volumes $R(t)$. Given any fixed region D, there is, at any moment in time, a material volume which coincides with it. Hence, for any fixed region D,

$$(1.7) \qquad\qquad \int_D \{c_t + \boldsymbol{\nabla} \cdot (c\boldsymbol{u})\} \, d\boldsymbol{x} = 0$$

for all t. This implies the pointwise PDE

$$(1.8) \qquad\qquad c_t + \boldsymbol{\nabla} \cdot (c\boldsymbol{u}) = 0.$$

Equation (1.8) applies in the fluid case with $c = \rho$, the density field.

Problem 1.5 (Density carried by incompressible flow).   A flow $\boldsymbol{u}(\boldsymbol{x}, t)$ is *incompressible* if each of its material regions has volume independent of time. Let $c(\boldsymbol{x}, t)$ be a density convected by such an incompressible flow, and let $S(t)$ be a material surface, that is, a time sequence of images of a given, fixed surface under the flow map. Show that if $S(0)$ is a level surface of $c(x, 0)$, then $S(t)$ is a level surface of $c(\boldsymbol{x}, t)$.

Solution. Let $\boldsymbol{y}_1$ and $\boldsymbol{y}_2$ be any two points on $S(t)$. There are $\boldsymbol{x}_1$ and $\boldsymbol{x}_2$ on $S(0)$ such that $\boldsymbol{y}_1 = \boldsymbol{X}(\boldsymbol{x}_1, t)$ and $\boldsymbol{y}_2 = \boldsymbol{X}(\boldsymbol{x}_2, t)$. Since $S(0)$ is a level surface of $c(x, 0)$, $c(\boldsymbol{x}_1, 0) = c(\boldsymbol{x}_2, 0)$. Now look at $c_1(t) := c(\boldsymbol{X}(\boldsymbol{x}_1, t), t)$. Compute

$$(1.5\text{-}1) \qquad \begin{aligned} \dot{c}_1(t) &= (c_t + \boldsymbol{X}_t(\boldsymbol{x}_1, t) \cdot \boldsymbol{\nabla} c)(\boldsymbol{X}(\boldsymbol{x}_1, t), t) \\ &= (c_t + \boldsymbol{u} \cdot \boldsymbol{\nabla} c)(\boldsymbol{X}(\boldsymbol{x}_1, t), t). \end{aligned}$$

Here, we have used $\boldsymbol{X}_t = \boldsymbol{u}(\boldsymbol{X}(\boldsymbol{x}_1, t), t)$. For incompressible flow, $\boldsymbol{\nabla} \cdot \boldsymbol{u} \equiv 0$, and the convection PDE reduces to $c_t + \boldsymbol{u} \cdot \boldsymbol{\nabla} c = 0$ for all $\boldsymbol{x}, t$. Hence, the right-hand side of (1.5-1) is zero, i.e., $c_1(t)$ is a constant independent of t, and $c(\boldsymbol{x}_1, 0) = c_1(0) = c_1(t) = c(\boldsymbol{y}_1, t)$. Similarly, $c(\boldsymbol{x}_2, 0) = c(\boldsymbol{y}_2, t)$. Since $c(\boldsymbol{x}_1, 0) = c(\boldsymbol{x}_2, 0)$, it follows that $c(\boldsymbol{y}_1, t) = c(\boldsymbol{y}_2, t)$. Since $\boldsymbol{y}_1$ and $\boldsymbol{y}_2$ are arbitrary points on $S(t)$, $S(t)$ is a level surface of $c(\boldsymbol{x}, t)$.

Problem 1.6 (Transformation of velocities as a flow map).   There is a swarm of stars. Let u denote their x-velocities as seen in a given (inertial) frame of reference. Their velocities v seen in an inertial frame with x-velocity U relative to the first inertial frame is

(1.6-1)
$$v = \frac{u - U}{1 - \frac{Uu}{c^2}}.$$

This is a formula from special relativity: c is the speed of light, and we have $|U|$ and all the $|u|$'s less than c. Then it follows from (1.6-1) that the $|v|$'s are all less than c too. Let $\rho(v, U)$ denote the distribution of these v's as seen in the moving frame, i.e., the number of stars with $v_1 < v < v_2$ is $\int_{v_1}^{v_2} \rho(v, U)dv$.

a) Show that $\rho(v, U)$ satisfies a convection PDE, with v as the "space-like" coordinate and U as "time".

b) Determine the solution of the convection PDE which is independent of U. This corresponds to an ensemble of stars whose x-velocity distribution looks the same in all inertial frames moving in the x-direction.

Solution.

a) Think of $u \to v(u, U) := \frac{u-U}{1-\frac{Uu}{c^2}}$ as a "flow map": Think of U as "time" and the star velocity v as "position" so that $v(u, U)$ is position at time U given that position at time 0 was u. The "velocity" at "time" U is

(1.6-2)
$$v_U(u, U) = -\frac{1 - \frac{u^2}{c^2}}{\left(1 - \frac{Uu}{c^2}\right)^2}.$$

Expressing this as a function of v gives the "velocity field" of the flow map (i.e., "velocity" as a function of "position" v). Solve $v = \frac{u-U}{1-\frac{Uu}{c^2}}$ for u to get $u = u(v, U) = \frac{v+U}{1+\frac{Uv}{c^2}}$, and substitute this into (1.6-2). The result is

$$V(v, U) := v_U(u(v, U), U) = -\frac{1 - \frac{v^2}{c^2}}{1 - \frac{U^2}{c^2}},$$

so the convection PDE with U, v and $V(v, U)$ in the roles of "time", "position" and "velocity" is

$$\rho_U - \frac{1}{1 - \frac{U^2}{c^2}} \partial_v \left\{ \left(1 - \frac{v^2}{c^2}\right)\rho \right\} = 0.$$

b) $\rho_U = 0$ implies that $(1 - \frac{v^2}{c^2})\rho$ is a uniform constant independent of v; so

$$\rho(v) = \frac{\rho(0)}{1 - \frac{v^2}{c^2}}.$$

Notice that this distribution is non-integrable on $-c < v < c$.

Convective flux and boundary conditions. Convective flux is a notion which arises from the conservation identity (1.7) with respect to a *fixed* region D. By using the divergence theorem, convert (1.7) into

(1.9) $$\frac{d}{dt}\int_D c\,d\boldsymbol{x} = -\int_{\partial D} c\boldsymbol{u}\cdot\boldsymbol{n}\,da.$$

If you think of $c(\boldsymbol{x},t)$ as the density of "stuff", (1.9) says that the rate of stuff entering D is minus the surface integral of the *convective flux* $\boldsymbol{f} = c\boldsymbol{u}$. Figure 1.3 is a cartoon explaining the right-hand side of (1.9). In time increment dt, the stuff inside the cylinder of volume $-(\boldsymbol{u}\cdot\boldsymbol{n}\,da)\,dt$ is carried inside D by the flow $\boldsymbol{u}$, contributing $-(c\boldsymbol{u}\cdot\boldsymbol{n})\,da\,dt$ to $\int_D c\,d\boldsymbol{x}$. The rate of contribution due to the union of patches da entirely covering ∂D is the surface integral on the right-hand side of (1.9).

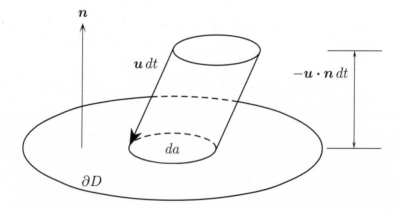

Figure 1.3.

Convective flux often appears in boundary conditions associated with the PDE (1.8). For instance, suppose that the velocity field $\boldsymbol{u}(\boldsymbol{x},t)$ has a jump discontinuity on a fixed $(n-1)$-dimensional surface in $\mathbb{R}^n$. Figure 1.4 depicts a two-dimensional surface S in $\mathbb{R}^3$ and particle paths (i.e., solutions of the ODE $\dot{\boldsymbol{x}}(t) = \boldsymbol{u}(\boldsymbol{x}(t),t)$) which have "kinks" as they cross S. We

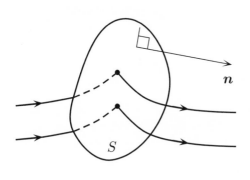

Figure 1.4.

assume that the trajectories really do cross S, instead of converging to S from both sides. Let $(c\boldsymbol{u})^+$ and $(c\boldsymbol{u})^-$ denote the right-hand and left-hand limits of $c\boldsymbol{u}$.[1] Stuff crosses S from the left at rate $(c\boldsymbol{u})^- \cdot \boldsymbol{n}$ per unit area, and emerges on the right at rate $(c\boldsymbol{u})^+ \cdot \boldsymbol{n}$ per unit area. Assuming that the surface S has no capacity to hold a surface density (i.e., nonzero stuff per unit area on S), the left and right rates must be equal, and so

$$(1.10) \qquad [c\boldsymbol{u}] \cdot \boldsymbol{n} = 0,$$

where $[\ ]$ denotes the right-hand limit minus the left-hand limit of the enclosed quantity, $c\boldsymbol{u}$ in this case. Notice that if the normal component of velocity jumps, i.e., $[\boldsymbol{u} \cdot \boldsymbol{n}] \neq 0$, then c jumps too: If cars slow down at the point on the highway where the highway patrol waits, then clearly the car density downstream from the highway patrol will be greater than the density upstream.

Here is another example of a boundary condition based on convective fluxes.

Elastic rebounds from a moving wall. Consider an ensemble of $N \gg 1$ identical systems: Each consists of a single particle moving along a line. At any time t, the particle is confined to the half line $x \leq \ell(t)$. In $x < \ell(t)$ the particle is free, and its position $x(t)$ and velocity $v(t) := \dot{x}(t)$ satisfy the ODEs

$$(1.11) \qquad \dot{x} = v, \quad \dot{v} = 0.$$

$x = \ell(t)$ represents a moving, solid wall. When a particle collides with this wall, it rebounds elastically: One must have $v > \dot{\ell}(t)$ just before collision, so that the incoming relative velocity $v - \dot{\ell}(t)$ is positive. The rebound relative velocity is $\dot{\ell}(t) - v$ and the rebound velocity (in the original frame of reference) is $2\dot{\ell} - v$.

The ensemble is described by an ensemble density function $n(x, v, t)$: The fraction of ensemble members in any region R of the (x, v) *phase plane* is $\int_R n(x, v, t)dx\, dv$. Now, we would really like to consider the density of $N \gg 1$ particles in the half space of $\mathbb{R}^3$ with $x < \ell(t)$. If the particles are

---

[1] Let $\boldsymbol{p}$ be any point on S. The right-hand (left-hand) limit of a function $f(\boldsymbol{x})$ as $\boldsymbol{x} \to \boldsymbol{p}$ is the limit of $f(\boldsymbol{p} + \boldsymbol{h})$ as $\boldsymbol{h} \to 0$ with $\boldsymbol{h} \cdot \boldsymbol{n} > (<)\, 0$.

so small that particle-particle collisions are rare, then the actual particle density in the space of x-positions and x-velocities evolves in the same way as the ensemble density. The ODEs (1.11) imply that the ensemble density $n(x, v, t)$ is convected by the velocity field

$$(v, 0)$$
$$\uparrow \quad \uparrow$$
$$x \; v \text{ components.}$$

Hence, $n(x, v, t)$ satisfies the convection PDE

(1.12) $$n_t + (vn)_x + (0n)_v = n_t + vn_x = 0$$

in $x < \ell(t)$.

The kinematics of particles rebounding from the wall is embodied in a boundary condition on n along the line $x = \ell(t)$ in the (x, v) plane. To derive the boundary condition, it is instructive to examine the "absorption" and "emission" of particles from the wall. The constant number of particles is expressed by the integral identity

$$\frac{d}{dt} \int_{-\infty}^{\infty} \int_{-\infty}^{\ell} n(x, v, t) dx \, dv = \int_{-\infty}^{\infty} \left\{ \int_{-\infty}^{\ell} n_t \, dx + \dot{\ell} n(\ell, v, t) \right\} dv = 0.$$

Substituting $n_t = -vn_x$ from (1.12), we obtain

(1.13) $$\int_{-\infty}^{\infty} (\dot{\ell} - v) n(\ell, v, t) dv = 0.$$

The integrand of (1.13) has a clear meaning: If $v > \dot{\ell}$, it is negative. Evidently, particles with velocities between $v > \dot{\ell}$ and $v + dv$ are "absorbed" upon collision with the wall at rate

(1.14) $$(v - \dot{\ell}) n(\ell, v, t) dv.$$

These particles rebound with velocities between $2\dot{\ell} - v - dv$ and $2\dot{\ell} - v$, and the rate of rebounds is (1.14) with v replaced by $2\dot{\ell} - v$ and dv by $-dv$, that is,

(1.15) $$(v - \dot{\ell}) n(\ell, 2\dot{\ell} - v, t) dv.$$

By conservation of particles, the rates (1.14) and (1.15) of collisions and rebounds are equal, so we have the boundary condition

(1.16) $$n(\ell, v, t) = n(\ell, 2\dot{\ell} - v, t).$$

Problems 1.7–1.10 treat the adiabatic expansion of particles in a box. Here, adiabatic means that the sides of the box move much more slowly than the particles inside.

Problem 1.7. We have an ensemble of $N \gg 1$ systems, each consisting of a particle moving along a line segment $-\frac{\ell(t)}{2} < x < \frac{\ell(t)}{2}$ (our box). The particle is free inside the box, and rebounds elastically off the walls $x = \frac{\ell(t)}{2}$ and $x = -\frac{\ell(t)}{2}$. Formulate the boundary value problem for the ensemble density $n(x, v, t)$. Show that this boundary value problem is invariant under simultaneous replacements of x by $-x$ and v by $-v$. (This being the case, there are solutions with $n(x, v, t) = n(-x, -v, t)$, and we will concentrate on these solutions in the remaining problems, 1.8 and 1.9.)

Solution. The ensemble density $n(x, v, t)$ satisfies the PDE

$$n_t + v n_x = 0$$

in $-\frac{\ell(t)}{2} < x < \frac{\ell(t)}{2}$, $-\infty < v < \infty$. The boundary conditions at the walls are

$$n(-\tfrac{\ell}{2}, v, t) = n(-\tfrac{\ell}{2}, -\dot{\ell} - v, t),$$

$$n(\tfrac{\ell}{2}, v, t) = n(\tfrac{\ell}{2}, \dot{\ell} - v, t).$$

The invariance of the PDE under reversal of the signs of x and v is obvious. Let us do the sign reversals in the first boundary condition. We get

$$n(\tfrac{\ell}{2}, -v, t) = n(\tfrac{\ell}{2}, \dot{\ell} + v, t).$$

Since this holds for all v, we can replace v by $-v$, and this gives the second boundary condition. Similarly, the sign reversals applied to the second boundary condition give the first.

Problem 1.8. We want to determine approximate solutions of the boundary value problem in Problem 1.7 in the *adiabatic limit* of slowly moving walls. A preliminary integral identity is crucial. Show that in the limit $\dot{\ell} \to 0$,

$$\int_{-\frac{\ell}{2}}^{\frac{\ell}{2}} n_t \, dx - \dot{\ell} v n_v(\tfrac{\ell}{2}, v, t) = O(\dot{\ell}^2).$$

Remember that we are working in the class of solutions with $n(x, v, t) = n(-x, -v, t)$.

Solution. The x-integral of the PDE over $-\frac{\ell}{2} < x < \frac{\ell}{2}$ gives

(1.8-1)
$$\int_{-\frac{\ell}{2}}^{\frac{\ell}{2}} n_t \, dx + v\{n(\tfrac{\ell}{2}, v, t) - n(-\tfrac{\ell}{2}, v, t)\} = 0.$$

By the boundary condition at $x = \frac{\ell}{2}$ and the $n(x, v, t) = n(-x, -v, t)$ symmetry, we have

$$n(-\tfrac{\ell}{2}, v, t) = n(-\tfrac{\ell}{2}, -\dot\ell - v, t) = n(\tfrac{\ell}{2}, \dot\ell + v, t),$$

and (1.8-1) becomes

$$\int_{-\frac{\ell}{2}}^{\frac{\ell}{2}} n_t \, dx + v\{n(\tfrac{\ell}{2}, v, t) - n(\tfrac{\ell}{2}, v + \dot\ell, t)\} = 0.$$

The Taylor expansion of $n(\frac{\ell}{2}, v + \dot\ell, t)$ in $\dot\ell$ gives

$$\int_{-\frac{\ell}{2}}^{\frac{\ell}{2}} n_t \, dx - \dot\ell v n_v(\tfrac{\ell}{2}, v, t) = O(\dot\ell^2).$$

Problem 1.9. In the adiabatic limit, we expect that there are solutions for $n(x, v, t)$ with weak dependence on x. We make the heuristic approximation $n \approx n(v, t)$ independent of x. The *velocity distribution*, defined by

$$\rho(v, t) := \int_{-\frac{\ell}{2}}^{\frac{\ell}{2}} n \, dx,$$

is approximated by $\rho(v, t) \approx \ell(t) n(v, t)$. Derive an approximate convection PDE in v-space for $\rho(v, t)$. Determine the flow map of v-space into itself.

Solution. Assuming n to be independent of x, the integral identity of Problem 1.8 reduces to

$$n_t(v, t) - \frac{\dot\ell(t)}{\ell(t)} v n_v(v, t) = 0,$$

where we have dropped the $O(\dot\ell^2)$ from the right-hand side. We convert this into a PDE for the velocity distribution by substituting $n = \frac{\rho}{\ell}$. We get

$$\rho_t + \left(-\frac{\dot\ell}{\ell} v \rho\right)_v = 0.$$

The "velocity field" in v-space is

$$u = -\frac{\dot\ell}{\ell} v.$$

Hence, the flow map $v \to V(v,t)$ is the solution of the initial value problem

$$
\begin{aligned}
V_t &= -\frac{\dot\ell}{\ell} V \quad \text{in } t > 0,\\
V(v,0) &= v.
\end{aligned}
$$

The solution is $V(v,t) = \frac{\ell(0)v}{\ell(t)}$, which implies that

$$
\ell(t)V(v,t) = \ell(0)v = \text{constant}.
$$

Problem 1.10. The flow map in v-space has a nice interpretation within classical mechanics. Assume that the speed of the particle is large compared to the speed of the walls. The "orbit" of the particle in the (x,v) phase plane is approximated by an oriented rectangle, depicted in Figure 1.5. The horizontal segments represent the particle moving back and forth with speed V between the walls, and the vertical segments, the (instantaneous) collisions which reverse the direction of motion. As the wall moves slowly, the collisions do work on the particle, and the speed V changes in response to the change in the width ℓ of the box. Although the actual changes in speed are discrete, we will approximate the time dependence of V due to the cumulative effect of many collisions by a smooth function $V(t)$. Derive an approximate ODE for $V(t)$, and show that it predicts that the area of the rectangular orbit in the (x,v) phase plane is a constant independent of time.

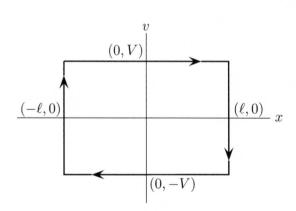

Figure 1.5.

Solution. One "orbit" around the (x,v) rectangle takes time $\Delta t = \frac{2\ell}{V}$, and in that time, there is a collision with the right wall $x = \frac{\ell}{2}$, which changes the velocity from V to $\dot\ell - V$, and a second collision with the left wall $x = \frac{-\ell}{2}$, which changes the velocity $\dot\ell - V$ to $-\dot\ell - (\dot\ell - V) = -2\dot\ell + V$. So the change

ΔV of speed in time Δt is $-2\dot{\ell}$. Hence, we propose the approximate ODE

$$\dot{V} = \frac{\Delta V}{\Delta t} = \frac{-2\dot{\ell}}{\left(\frac{2\ell}{V}\right)} = -\frac{\dot{\ell}}{\ell}V.$$

This is the ODE of the flow map in Problem 1.8, and it implies that $V(t)\ell(t)$ is a constant independent of t.

Convective derivative. Let $\boldsymbol{x}(t)$ be any material point in a flow $\boldsymbol{u}(\boldsymbol{x},t)$, so

$$\text{(1.17)} \qquad\qquad \dot{\boldsymbol{x}}(t) = \boldsymbol{u}(\boldsymbol{x}(t),t).$$

The time derivative of a function $f(\boldsymbol{x},t)$ seen at the material point $\boldsymbol{x}(t)$ is, by the chain rule,

$$\text{(1.18)} \qquad\qquad \frac{d}{dt}f(\boldsymbol{x}(t),t) = (f_t + \boldsymbol{u}\cdot\boldsymbol{\nabla}f)(\boldsymbol{x}(t),t).$$

The directional derivative of $f(\boldsymbol{x},t)$ in $(\boldsymbol{x},t)$ spacetime, which appears on the right-hand side, is called the *convective derivative* of f,

$$\text{(1.19)} \qquad\qquad \frac{Df}{Dt} := f_t + \boldsymbol{u}\cdot\boldsymbol{\nabla}f.$$

The convective derivative informs an alternative derivation of the PDE (1.8) for the density c carried in a flow $\boldsymbol{u}$. Let us start with the conservation identity (1.3) which says that the total amount of stuff in material region $R(t)$ is a constant independent of t. In the limit of $R(t)$ shrinking to the single material point $\boldsymbol{x}(t)$, (1.3) reduces to

$$\text{(1.20)} \qquad\qquad \frac{1}{V(t)}\frac{d}{dt}(c(\boldsymbol{x}(t),t)V(t)) \to 0,$$

where $V(t)$ is the volume of $R(t)$. Carry out time differentiation of the product cV and use the chain rule and definition (1.19) of the convective derivative to rewrite (1.20) as

$$\text{(1.21)} \qquad\qquad \frac{Dc}{Dt}(\boldsymbol{x}(t),t) + \frac{\dot{V}(t)}{V(t)}c(\boldsymbol{x}(t),t) \to 0.$$

Now recall (1.6), which says that $\frac{\dot{V}}{V}$ converges to $(\boldsymbol{\nabla}\cdot\boldsymbol{u})(\boldsymbol{x}(t),t)$ as $R(t)$ shrinks to a point $\boldsymbol{x}(t)$. Hence, (1.6) and (1.21) lead to

$$\text{(1.22)} \qquad\qquad \left\{\frac{Dc}{Dt} + (\boldsymbol{\nabla}\cdot\boldsymbol{u})c\right\}(\boldsymbol{x}(t),t) = 0.$$

For any fixed point $(\boldsymbol{x}, t)$ in spacetime, there is a material point which matches the given $\boldsymbol{x}$ at the given t, so (1.22) holds for all fixed $(\boldsymbol{x}, t)$, i.e.,

$$(1.23) \qquad \frac{Dc}{Dt} + (\boldsymbol{\nabla} \cdot \boldsymbol{u})c = 0.$$

Rewrite (1.23) as

$$c_t + \boldsymbol{u} \cdot \boldsymbol{\nabla} c + (\boldsymbol{\nabla} \cdot \boldsymbol{u})c = 0$$

and use the product rule identity

$$\boldsymbol{u} \cdot \boldsymbol{\nabla} c + (\boldsymbol{\nabla} \cdot \boldsymbol{u})c = \boldsymbol{\nabla} \cdot (c\boldsymbol{u})$$

to see the equivalence between (1.8) and (1.23).

Convected scalars. The real-valued function $f(\boldsymbol{x}, t)$ is called a *convected scalar* in the flow $\boldsymbol{u}$ if

$$(1.24) \qquad \frac{Df}{Dt} = f_t + (\boldsymbol{u} \cdot \boldsymbol{\nabla})f = 0.$$

Unlike a convected density, there is no adjustment in its value at a material point due to local expansion or concentration of material regions. Convected scalars *do* arise in the real, common-sense world. For instance, suppose two kinds of particles A and B are carried in the flow $\boldsymbol{u}$, so that we have two convected densities c_A and c_B. The fraction of A particles is given by

$$(1.25) \qquad f_A := \frac{c_A}{c_A + c_B}$$

and similarly for the fraction of B particles. These fractions are convected scalars, as we will show in Problem 1.11. In ideal gas dynamics with no dissipative process, the *entropy per unit mass* is a convected scalar. In the special case of *incompressible* flow (in which all material regions have volumes independent of t), it follows that $\boldsymbol{\nabla} \cdot \boldsymbol{u} = 0$. Hence, (1.23) reduces to $\frac{Dc}{Dt} = 0$, and we see that a convected density in incompressible flow is a convected scalar as well.

Problem 1.11 (A convected scalar). Let c_A and c_B be convected densities of "A" and "B" particles in a flow $\boldsymbol{u}$. Show that the fraction of A particles as defined in (1.25) is a convected scalar.

Solution. Logarithmic time differentiation of (1.25) at a material point gives

$$(1.11\text{-}1) \qquad \frac{1}{f_A}\frac{Df_A}{Dt} = \frac{1}{c_A}\frac{Dc_A}{Dt} - \frac{1}{c_A+c_B}\frac{D}{Dt}(c_A+c_B).$$

Now substitute

$$\frac{1}{c_A}\frac{Dc_A}{Dt} = \frac{1}{c_B}\frac{Dc_B}{Dt} = -\boldsymbol{\nabla}\cdot\boldsymbol{u},$$

and the right-hand side of (1.11-1) becomes

$$-\boldsymbol{\nabla}\cdot\boldsymbol{u} + \boldsymbol{\nabla}\cdot\boldsymbol{u} = 0.$$

Gradient of a convected scalar. The notion of convective derivative leads to a clear analysis of how the *gradient* of a convected scalar or density evolves. For simplicity, consider a convected scalar which satisfies (1.24), i.e.,

$$(1.26) \qquad \frac{Df}{Dt} = \partial_t f + u_j \partial_j f = 0.$$

Here, u_j are the components of the velocity field $\boldsymbol{u}$ and $\partial_j c$ is the partial derivative of c with respect to x_j. The twice repeated index j in (1.26) means summation over $j = 1,\ldots,n$. Compute the ith partial derivative of (1.26),

$$\partial_t(\partial_i f) + u_j\partial_j(\partial_i f) = -(\partial_i u_j)\partial_j f,$$

or

$$(1.27) \qquad \frac{D(\partial_i f)}{Dt} = -(\partial_i u_j)\partial_j f, \quad i = 1,\ldots,n.$$

Equation (1.27) holds for all events in $(\boldsymbol{x},t)$ spacetime, and in particular along the world line of a material point $\boldsymbol{x}(t)$. Along this world line, $\partial_i f$ are values at events $(\boldsymbol{x}(t),t)$ and, as such, functions of time. Likewise for the velocity field derivatives $\partial_i u_j$. The left-hand side is the time derivative of $(\partial_i f)(\boldsymbol{x}(t),t)$. Hence, (1.27) is a system of ODEs for the values of $\partial_1 f,\ldots,\partial_n f$ at $(\boldsymbol{x}(t),t)$.

Example. In polar coordinates (r,ϑ) of the plane, the flow $\boldsymbol{u}$ due to a point vortex in incompressible two-dimensional fluid is

$$\boldsymbol{u} = \frac{\Gamma}{2\pi r}\,\boldsymbol{e}_\vartheta.$$

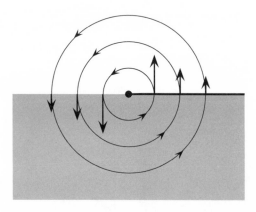

Figure 1.6.

The constant Γ measures the strength of the vortex. Figure 1.6 depicts the circular orbits of material points in this vortex flow. The vertical arrows represent the restriction of $\boldsymbol{u}$ to the x-axis. Let $c(r, \vartheta, t)$ be a convected density in this vortex flow. Since the vortex flow is incompressible, c is a convected scalar with $\frac{Dc}{Dt} = c_t + \boldsymbol{u} \cdot \boldsymbol{\nabla} c = 0$. The polar coordinate form of this PDE is

$$c_t + \left(0 \boldsymbol{e}_r + \frac{\Gamma}{2\pi r} \, \boldsymbol{e}_\vartheta \right) \cdot \left(c_r \boldsymbol{e}_r + \frac{1}{r} c_\vartheta \boldsymbol{e}_\vartheta \right) = 0,$$

or

$$(1.28) \qquad \frac{Dc}{Dt} = c_t + \frac{\Gamma}{2\pi r^2} \, c_\vartheta = 0.$$

We want to track the gradient $\boldsymbol{\nabla} c = c_r \boldsymbol{e}_r + \frac{1}{r} c_\vartheta \boldsymbol{e}_\vartheta$ along the path of a material point. Hence, we need evaluations of c_r and c_ϑ. Differentiation of (1.28) with respect to r and ϑ gives

$$(1.29) \qquad \frac{D(c_r)}{Dt} = (c_r)_t + \frac{\Gamma}{2\pi r^2} (c_r)_\vartheta = \frac{\Gamma}{\pi r^3} c_\vartheta,$$

$$(1.30) \qquad \frac{D(c_\vartheta)}{Dt} = (c_\vartheta)_t + \frac{\Gamma}{2\pi r^2} (c_\vartheta)_\vartheta = 0.$$

Now (1.29) and (1.30) are special cases of the ODE (1.27) in polar coordinates. They give the time evolutions of c_r and c_ϑ at a material point whose polar coordinates $r(t), \vartheta(t)$ satisfy

$$(1.31) \qquad \dot{r} = 0, \quad \dot{\vartheta} = \frac{\Gamma}{2\pi r^2}.$$

From (1.31) we see that material points move in circles of various radii r, with uniform angular velocities $\frac{\Gamma}{2\pi r^2}$. From the ODEs (1.29) and (1.30) it follows that at any such material point, $c_\vartheta = \gamma$, a constant independent of

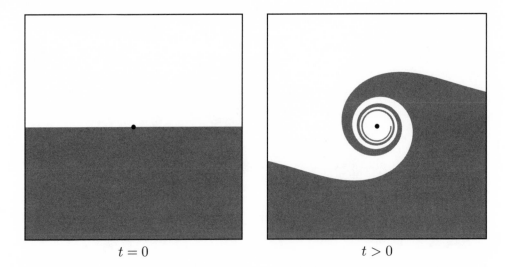

$$t = 0 \qquad\qquad\qquad t > 0$$

Figure 1.7.

time, and then

$$\frac{D(c_r)}{Dt} = \frac{\Gamma\gamma}{\pi r^3}.$$

A material point has $r(t) = r = $ constant, so the right-hand side is independent of time. Thus, the value of c_r seen at a fluid particle circulating around the origin grows indefinitely at a uniform rate. Figure 1.7 gives a visual sense of this growth in c_r. In the first panel, the shading represents stuff in the lower half plane at time $t = 0$. The second panel shows the stuff after it has been stirred by the vortex flow for time $t > 0$. There are spiral bands of stuff which become very fine and nearly circular as $t \to \infty$. Points of high and low concentration become very close to each other, and this indicates the growth of c_r.

Problem 1.12 (Flows that preserve gradients).

a) Let $c(\boldsymbol{x}, t)$ be a density convected in an incompressible flow $\boldsymbol{u}(\boldsymbol{x}, t)$ in $\mathbb{R}^n$. Show that $\frac{D}{Dt}|\boldsymbol{\nabla} c|^2 = -2\boldsymbol{\nabla} c \cdot (S\boldsymbol{\nabla} c)$, where S is the *strain matrix* of the flow $\boldsymbol{u}$, with cartesian components $S_{ij} := \frac{1}{2}(\partial_i u_j + \partial_j u_i)$.

b) What is the most general incompressible flow for which $|\boldsymbol{\nabla} c|^2$ is always time-independent at material points?

Solution.

a) We calculate, with the help of (1.27),

$$\frac{D}{Dt}|\boldsymbol{\nabla}c|^2 \;=\; \frac{D}{Dt}(\partial_i c\,\partial_i c)$$

$$=\; 2\partial_i c\frac{D}{Dt}\partial_i c = -2\partial_i c(\partial_i u_j)\partial_j c.$$

The double summation (over i,j) is invariant under interchange of i and j. Hence,

$$\frac{D}{Dt}|\boldsymbol{\nabla}c|^2 \;=\; -2\partial_i c\Big\{\frac{1}{2}(\partial_i u_j + \partial_j u_i)\Big\}\partial_j c$$

$$=\; -2\partial_i c S_{ij}\partial_j c$$

$$=\; -2\boldsymbol{\nabla}c\cdot(S\boldsymbol{\nabla}c).$$

b) For arbitrary $\boldsymbol{v} := \boldsymbol{\nabla}c$ we require $\boldsymbol{v}\cdot(S\boldsymbol{v}) = 0$. First, take $\boldsymbol{v} = \boldsymbol{e}_i$, the ith unit vector with components

$$(\boldsymbol{e}_i)_j = \delta_{ij} := \left[\begin{array}{ll} 1, & i = j, \\ 0, & i \neq j. \end{array}\right.$$

We find that $0 = \boldsymbol{e}_i\cdot(S\boldsymbol{e}_i) = S_{ii} = \partial_i u_i$ (with no summation over i); so for each i, u_i is *independent of* x_i. Next, take $\boldsymbol{v} = \boldsymbol{e}_i + \boldsymbol{e}_j$, where $\boldsymbol{e}_i \neq \boldsymbol{e}_j$. We have $0 = (\boldsymbol{e}_i + \boldsymbol{e}_j)\cdot S(\boldsymbol{e}_i + \boldsymbol{e}_j) = S_{ii} + S_{ij} + S_{ji} + S_{jj} = 2S_{ij} = \partial_i u_j + \partial_j u_i$, so

(1.12-1) $$\partial_j u_i = -\partial_i u_j.$$

In the right-hand side, $\partial_i u_j$ is independent of x_j, and it follows that u_i is a linear combination of the x_j with $j \neq i$, plus an additive constant. In mathematical notation,

(1.12-2) $$u_i = \sum_{j\neq i} w_{ij}x_j + v_i,$$

where the w_{ij} and v_i are constants. Substituting (1.12-2) into (1.12-1), we find that $w_{ij} = -w_{ji}$. The summary of equation (1.12-2) in vector notation is

$$\boldsymbol{u} = W\boldsymbol{x} + \boldsymbol{v},$$

where W is the antisymmetric matrix with components w_{ij} and $\boldsymbol{v}$ is the constant vector with components v_i. Hence, the flow $\boldsymbol{u}$ is a sum of solid body rotation ($W\boldsymbol{x}$) and translation at uniform velocity $\boldsymbol{v}$.

Problem 1.13 (Vector analog of a convected scalar).   First, the analog is *not* $\frac{D}{Dt}\boldsymbol{v} = 0$. Instead, paint little oriented line segments in cookie dough to represent the initial values of a vector field; then deform the dough, and the induced displacement, rotation and stretching of the line segments will represent "convection" of the original vector field. Here is the mathematical formulation (Figure 1.8). Let $\boldsymbol{u}(\boldsymbol{x}, t)$ be a given velocity field and $\boldsymbol{v}(\boldsymbol{x}, t)$ another time-dependent vector field with a specific connection to $\boldsymbol{u}$. Let $C : \boldsymbol{x}(s, t)$ be a material curve of the flow $\boldsymbol{u}$, with s labeling material points. If at any one time t the restriction of $\boldsymbol{v}$ to C is the tangent vector $\boldsymbol{\tau}(s, t) := \boldsymbol{x}_s(s, t)$, then $\boldsymbol{v}$ restricted to C is the tangent vector $\boldsymbol{\tau}(s, t)$ for all time. In this case, we say that the *vector field* $\boldsymbol{v}(\boldsymbol{x}, t)$ is *convected by the flow* $\boldsymbol{u}$.

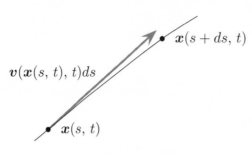

Figure 1.8.

We want to derive an evolution PDE for $\boldsymbol{v}(\boldsymbol{x}, t)$.

a) First, a little geometric preamble: Let $\boldsymbol{x}(s, t)$ be a parametric representation of a material curve in a flow $\boldsymbol{u}(\boldsymbol{x}, t)$, such that s labels material points. Show that the tangent vector $\boldsymbol{\tau}(s, t) := \boldsymbol{x}_s(s, t)$ satisfies

(1.13-1) $$\boldsymbol{\tau}_t = (\boldsymbol{\tau} \cdot \boldsymbol{\nabla})\boldsymbol{u}.$$

b) Show that

(1.13-2) $$\boldsymbol{v}_t + (\boldsymbol{u} \cdot \boldsymbol{\nabla})\boldsymbol{v} = (\boldsymbol{v} \cdot \boldsymbol{\nabla})\boldsymbol{u}.$$

Solution.

a) Since fixed values of s label material points, $\boldsymbol{x}(s, t)$ satisfies

$$\boldsymbol{x}_t(s, t) = \boldsymbol{u}(\boldsymbol{x}(s, t), t),$$

or, in component notation,

$$\partial_t x_i = u_i(\boldsymbol{x}(s, t), t).$$

Differentiation with respect to s gives

$$\partial_t(\partial_s x_i)(s, t) = (\partial_s x_j(s, t))\partial_j u_i(\boldsymbol{x}(s, t), t),$$

or

$$\partial_t \tau_i(s, t) = ((\tau_j(s, t)\partial_j)u_i)(\boldsymbol{x}(s, t), t)$$

with summation over the repeated index j. In vector notation, $\boldsymbol{\tau}_t = (\boldsymbol{\tau} \cdot \boldsymbol{\nabla})\boldsymbol{u}$.

b) Substitute $\boldsymbol{\tau}(s,t) = \boldsymbol{v}(\boldsymbol{x}(s,t),t)$ into (1.13-1) to get

$$\frac{D\boldsymbol{v}}{dt} = \boldsymbol{v}_t + (\boldsymbol{u} \cdot \boldsymbol{\nabla})\boldsymbol{v} = (\boldsymbol{v} \cdot \boldsymbol{\nabla})\boldsymbol{u}.$$

Problem 1.14 (Vector analogs of a convected density).   Recall that the volume integral of a *convected density* over any material region is conserved in time. There are two vector analogs in spatial $\mathbb{R}^3$. First, in electrically conducting fluid, the magnetic flux, or surface integral of the magnetic field over any material surface, is conserved in time. So, in this analog, the surface integral of a vector field takes on the role of mass, and the vector field, whose integrals over material surfaces are conserved, takes on the role of density. Second, we can consider vector fields whose *line* integral over any material curve is conserved. We begin with this second analog and an example of it.

a) The vector field $\boldsymbol{g}(\boldsymbol{x},t)$ in $\mathbb{R}^3$ has

(1.14-1) $$\frac{d}{dt} \int_{C(t)} \boldsymbol{g}(\boldsymbol{x},t) \cdot d\boldsymbol{x} = 0$$

for any material curve $C(t)$. Give a class of vector fields which automatically satisfy equation (1.14-1). Derive an evolution PDE for $\boldsymbol{g}(\boldsymbol{x},t)$, first in *coordinate* notation. Next, convert your equations to coordinate-free vector notation. This makes it easier for you to see that the equation of $\boldsymbol{g}$ is *not* the same as that of a convected vector, as in Problem 1.13.

b) Now for the first analog: Let $\boldsymbol{\omega}(\boldsymbol{x},t)$ be a vector field on $\mathbb{R}^3$ whose surface integral

$$\int_{S(t)} \boldsymbol{\omega} \cdot \boldsymbol{n} \, da$$

over any material surface $S(t)$ is conserved in time. Derive the evolution PDE for $\boldsymbol{\omega}(\boldsymbol{x},t)$. Here is an outline of an argument that cuts to the essence and avoids vector identities that everybody looks up but nobody remembers. First, we consider regions which are "bundles" of integral curves of $\boldsymbol{\omega}$, as depicted in Figure 1.9. Argue that if a material region is such a bundle at

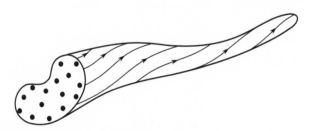

Figure 1.9.

one time, it is so for all times. Argue that integral curves of $\boldsymbol{\omega}$ are material curves. In this case, $\boldsymbol{\omega}$ is everywhere parallel to a convected vector $\boldsymbol{v}$ in the sense of Problem 1.13. You will have to come up with some idea for the local proportionality coefficient between $\boldsymbol{\omega}$ and $\boldsymbol{v}$.

Solution.

a) For any smooth function $f(\boldsymbol{x}, t)$,

$$\int_{C(t)} \boldsymbol{\nabla} f \cdot d\boldsymbol{x} = f(\boldsymbol{x}(1, t), t) - f(\boldsymbol{x}(0, t), t).$$

The right-hand side represents the difference of f values at the endpoints of $C(t)$. If $f(\boldsymbol{x}, t)$ is a convected scalar, the endpoint values are time-independent, so

$$\frac{d}{dt} \int_{C(t)} \boldsymbol{\nabla} f \cdot d\boldsymbol{x} = 0.$$

In summary, (1.14-1) is satisfied if $\boldsymbol{g}$ is a gradient of a convected scalar.

We derive the PDE for $\boldsymbol{g}(\boldsymbol{x}, t)$: Let $\boldsymbol{x}(s, t)$ be a parametric representation of $C(t)$ with s in $[0, 1]$ labeling material points. We rewrite (1.14-1) as

(1.14-2)
$$\begin{aligned} 0 &= \frac{d}{dt} \int_0^1 \boldsymbol{g}(\boldsymbol{x}(s, t), t) \cdot \boldsymbol{\tau}(s, t) \, ds \\ &= \int_0^1 \left\{ \frac{D\boldsymbol{g}}{Dt} \cdot \boldsymbol{\tau} + \boldsymbol{g} \cdot ((\boldsymbol{\tau} \cdot \boldsymbol{\nabla})\boldsymbol{u}) \right\} ds. \end{aligned}$$

Here, $\boldsymbol{\tau} := \boldsymbol{x}_s(s, t)$ is the tangent vector of $C(t)$ and the second equality uses $\boldsymbol{\tau}_t = (\boldsymbol{\tau} \cdot \boldsymbol{\nabla})\boldsymbol{u}$ as in (1.13-1). In component notation, (1.14-2) reads

$$\int_0^1 \left\{ \frac{Dg_i}{Dt} + g_j \partial_i u_j \right\} \tau_i \, ds = 0,$$

and our equations for the g_i are

(1.14-3)
$$\frac{Dg_i}{Dt} = -(\partial_i u_j)g_j,$$

where there is summation over the repeated index j.

We write these equations in vector notation. First, rewrite (1.14-3) as

(1.14-4)
$$\frac{D}{Dt}g_i + g_j \partial_j u_i = -g_j(\partial_i u_j - \partial_j u_i).$$

The left-hand side of (1.14-4) is the ith component of the vector

$$\frac{D\boldsymbol{g}}{Dt} + (\boldsymbol{g} \cdot \boldsymbol{\nabla})\boldsymbol{u}.$$

It remains to determine what vector has components as in the right-hand side. We simplify by reorienting our cartesian coordinates so that $\boldsymbol{e}_1$ is parallel to $\boldsymbol{g}$, and so $\boldsymbol{g} = g_1 \boldsymbol{e}_1$. Then the right-hand side of (1.14-4) reduces to

$$-g_1(\partial_i u_1 - \partial_1 u_i),$$

and the vector with these components is

(1.14-5) $\qquad -g_1\left\{(\partial_2 u_1 - \partial_1 u_2)\boldsymbol{e}_2 + (\partial_3 u_1 - \partial_1 u_3)\boldsymbol{e}_3\right\}.$

We recognize components of the *curl* of $\boldsymbol{u}$,

$$\begin{aligned}
\boldsymbol{\omega} := \boldsymbol{\nabla} \times \boldsymbol{u} &= w_1 \boldsymbol{e}_1 + w_2 \boldsymbol{e}_2 + w_3 \boldsymbol{e}_3 \\
&= (\partial_1 u_2 - \partial_2 u_1)\boldsymbol{e}_3 + (\partial_2 u_3 - \partial_3 u_2)\boldsymbol{e}_1 + (\partial_3 u_1 - \partial_1 u_3)\boldsymbol{e}_2.
\end{aligned}$$

(Memory aid: Write out the term $(\partial_1 u_2 - \partial_2 u_1)\boldsymbol{e}_3$ and then the other two are obtained by cyclic permutation of integers $1, 2, 3$.) Hence, (1.14-5) becomes

$$\begin{aligned}
-g_1\left\{-w_3 \boldsymbol{e}_2 + w_2 \boldsymbol{e}_3\right\} &= -g_1\left\{-w_3 \boldsymbol{e}_3 \times \boldsymbol{e}_1 + w_2 \boldsymbol{e}_1 \times \boldsymbol{e}_2\right\} \\
&= -g_1 \boldsymbol{e}_1 \times \left\{w_2 \boldsymbol{e}_2 + w_3 \boldsymbol{e}_3\right\} \\
&= -g_1 \boldsymbol{e}_1 \times \left\{w_1 \boldsymbol{e}_1 + w_2 \boldsymbol{e}_2 + w_3 \boldsymbol{e}_3\right\} \\
&= -\boldsymbol{g} \times \boldsymbol{\omega}.
\end{aligned}$$

(Another memory aid: $\boldsymbol{e}_1 \times \boldsymbol{e}_2 = \boldsymbol{e}_3$ and, by cyclic permutation, $\boldsymbol{e}_2 \times \boldsymbol{e}_3 = \boldsymbol{e}_1$ and $\boldsymbol{e}_3 \times \boldsymbol{e}_1 = \boldsymbol{e}_2$. Also, there is antisymmetry, $\boldsymbol{a} \times \boldsymbol{b} = -\boldsymbol{b} \times \boldsymbol{a}$, and $\boldsymbol{e}_1 \times \boldsymbol{e}_1 = 0$ as a special case.) In summary, the vector form of (1.14-4) is

(1.14-6) $\qquad \dfrac{D\boldsymbol{g}}{Dt} = -(\boldsymbol{g} \cdot \boldsymbol{\nabla})\boldsymbol{u} - \boldsymbol{g} \times \boldsymbol{\omega}.$

This is *not* the convected vector equation

(1.14-7) $\qquad \dfrac{D\boldsymbol{v}}{Dt} = (\boldsymbol{v} \cdot \boldsymbol{\nabla})\boldsymbol{u}.$

Why is (1.14-6) different from (1.14-7)? In elementary vector analysis, we are used to calling the gradient of a function "a vector field". But geometers (rightly) beg to differ: In its essence, the gradient is really a *linear function*, telling you the *change* in a scalar field when you make a small displacement. For geometers, the prototypical vector is *spatial displacement*, a different object, really. Take our streaks in cookie dough: When we stretch the dough in the direction parallel to the streaks, they *elongate*. That is why $+(\boldsymbol{v} \cdot \boldsymbol{\nabla})\boldsymbol{u}$ appears in the right-hand side of (1.14-7). On the other hand, suppose the sugar concentration is increasing in the direction of our streaks, so that the gradient "vector" is parallel to the streaks. When we stretch the dough, the gradient *decreases* and hence $-(\boldsymbol{g} \cdot \boldsymbol{\nabla})\boldsymbol{u}$ appears in (1.14-6).

b) Consider a material region $R(t)$ which at $t = 0$ is a bundle of integral curves of $\boldsymbol{\omega}(\boldsymbol{x}, 0)$, as depicted in Figure 1.9. The surface integral $\int_{S(t)} \boldsymbol{\omega} \cdot \boldsymbol{n}\, da$ through any material patch $S(t)$ on the walls is constant in time. At time zero, we have $\boldsymbol{\omega} \cdot \boldsymbol{n} = 0$ on the walls, so $\int_{S(0)} \boldsymbol{\omega} \cdot \boldsymbol{n}\, da = 0$. Hence, $\int_{S(t)} \boldsymbol{\omega} \cdot \boldsymbol{n}\, da = 0$ for all material patches on the walls, and $\boldsymbol{\omega} \cdot \boldsymbol{n} = 0$ on the walls for all time. The surface of $R(t)$ is made up of integral curves of $\boldsymbol{\omega}(\boldsymbol{x}, t)$, and the interior of $R(t)$ is filled with integral curves which don't break the surface. In summary, $R(t)$ is a bundle of integral curves for all time. We also note that the surface integral through material cross-sections, generally nonzero, is constant in time (but not necessarily the same for different cross-sections, unless we further assume $\boldsymbol{\nabla} \cdot \boldsymbol{\omega} = 0$). By considering the limit of such a bundle shrinking down to a material curve, we see that integral curves of $\boldsymbol{\omega}$ are also material curves.

Let $\boldsymbol{v}(\boldsymbol{x}, t)$ be the convected vector field, satisfying (1.14-7), whose initial values at $t = 0$ are $\boldsymbol{\omega}(\boldsymbol{x}, 0)$. Integral curves of $\boldsymbol{v}$ are material curves, and since they coincide with the integral curves of $\boldsymbol{\omega}$ at $t = 0$, they are the same as the integral curves of $\boldsymbol{\omega}$. Let $\boldsymbol{x}(s, t)$ be a parametric representation of one of these integral curves $C(t)$, with s labeling material points. We can choose the labeling s so that $\boldsymbol{v}(\boldsymbol{x}(s, 0), 0) = \boldsymbol{\tau}(s, 0) := \boldsymbol{x}_s(s, 0)$. By the definition of a convected vector as in Problem 1.13,

$$(1.14\text{-}8) \qquad\qquad \boldsymbol{v}(\boldsymbol{x}(s, t), t) = \boldsymbol{\tau}(s, t)$$

for all t. We also have

$$(1.14\text{-}9) \qquad\qquad \boldsymbol{\omega}(\boldsymbol{x}(s, t), t) = \lambda(s, t)\boldsymbol{v}(\boldsymbol{x}(s, t), t).$$

The local coefficient $\lambda(s, t)$ remains to be determined. Assuming for now that it is given, time differentiation of (1.14-9) and use of (1.13-2) gives

$$
\begin{aligned}
\frac{D\boldsymbol{\omega}}{Dt} &= \lambda_t \boldsymbol{v} + \lambda \frac{D\boldsymbol{v}}{Dt} \\
&= \lambda_t \boldsymbol{v} + \lambda (\boldsymbol{v} \cdot \boldsymbol{\nabla})\boldsymbol{u} \\
&= \frac{\lambda_t}{\lambda}\boldsymbol{\omega} + (\boldsymbol{\omega} \cdot \boldsymbol{\nabla})\boldsymbol{u}.
\end{aligned}
$$

(1.14-10)

Now we deal with λ: Consider a "thin" bundle of integral curves of $\boldsymbol{\omega}$ about $C(t)$ and, in particular, a short material section of it, depicted in Figure 1.10. The face of this oblique cylinder through $\boldsymbol{x}(s, t)$ has area da and unit normal $\boldsymbol{n}$. The geometric parameters da, $\boldsymbol{n}$, and $\boldsymbol{\tau}(s, t)$ (the displacement from $\boldsymbol{x}(x, t)$ to $\boldsymbol{x}(s + ds, t)$ is $\boldsymbol{\tau}(s, t)\, ds$) can vary in time, subject to a constraint: The volume of the material cylinder is $(\boldsymbol{\tau}\, ds) \cdot (\boldsymbol{n}\, da) = (\boldsymbol{\tau} \cdot \boldsymbol{n}\, da)\, ds$, and its time rate of change satisfies (1.6), that is,

$$(1.14\text{-}11) \qquad\qquad (\boldsymbol{\tau} \cdot \boldsymbol{n}\, da)_t = (\boldsymbol{\nabla} \cdot \boldsymbol{u})(\boldsymbol{\tau} \cdot \boldsymbol{n}\, da),$$

where $\boldsymbol{\nabla} \cdot \boldsymbol{u}$ is evaluated at $\boldsymbol{x} = \boldsymbol{x}(s, t)$ and we have dropped the ds, which is constant in time.

Next, the surface integral of $\boldsymbol{\omega}$ over the cross-section of the bundle through $\boldsymbol{x}(s, t)$, namely

$$\boldsymbol{\omega}(\boldsymbol{x}(s, t), t) \cdot \boldsymbol{n}\, da,$$

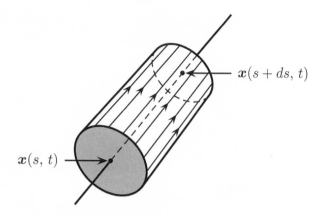

Figure 1.10.

is a constant independent of time. By (1.14-8) and (1.14-9),

$$\boldsymbol{\omega}(\boldsymbol{x}(s,t),t) = \lambda(s,t)\boldsymbol{\tau}(s,t),$$

so $\lambda\boldsymbol{\tau} \cdot \boldsymbol{n}\, da$ is a constant independent of time. Hence,

(1.14-12) $$\lambda_t \boldsymbol{\tau} \cdot \boldsymbol{n}\, da + \lambda(\boldsymbol{\tau} \cdot \boldsymbol{n}\, da)_t = 0,$$

and it follows from (1.14-11) and (1.14-12) that

$$\frac{\lambda_t}{\lambda} = -\boldsymbol{\nabla} \cdot \boldsymbol{u}.$$

Substituting this result for $\frac{\lambda_t}{\lambda}$ into (1.14-10), we obtain the evolution PDE for $\boldsymbol{\omega}$,

(1.14-13) $$\frac{D\boldsymbol{\omega}}{Dt} = (\boldsymbol{\omega} \cdot \boldsymbol{\nabla})\boldsymbol{u} - (\boldsymbol{\nabla} \cdot \boldsymbol{u})\boldsymbol{\omega}.$$

Geometric postscript. In the geometry world, densities are called *forms*. In $\mathbb{R}^3$, an ordinary density as the amount of "stuff" per unit volume is called a *three-form*. A "vector field" which determines values of surface integrals over two-dimensional surfaces is a *two-form*, and a "vector field" which determines line integrals over one-dimensional curves is a *one-form*. As such, their physical units contain $1 \div (\text{length})^3$, $1 \div (\text{length})^2$ and $1 \div \text{length}$, respectively. In particular, the $1 \div \text{length}$ unit of a one-form is a dead giveaway that it is not the same thing as a displacement with the unit of length. It is a coincidence that one- and two-forms in $\mathbb{R}^3$ are specified by three real numbers (their "components") so that in many "everyday" uses, one- and two-forms are "mapped" into vectors.

Guide to bibliography. Recommended references for this chapter are Batchelor [**2**], Chorin & Mardsen [**4**], Landau & Lifshitz [**17**], and Ockendon & Ockendon [**20**].

In all these references, analysis of the convection PDE is embedded in the bigger picture of fluid mechanics. Ockendon & Ockendon [**20**] base their (clear, elegant) derivations on moving material regions and the transport theorem. Chorin & Mardsen [**4**] also present the transport theorem. Batchelor's [**2**] discussion of transport into or from a material volume is based on heuristic vector analysis, but the reader needs to be self-sufficient in supplying the visual insight (such as Figure 1.2). Batchelor [**2**] emphasizes the local kinematics of a fluid about a material point, and he likes convective derivatives. Landau & Lifshitz [**17**] introduce and use the term "flux" as we do. This makes sense, because many of their derivations are based on fixed control volumes rather than moving material volumes.

Chapter 2

Diffusion

Brownian motion. Small particles in water are buffeted by impacts of surrounding water molecules. You can experience this condition on a New York subway platform, where you are the particle. Let us record displacements $\boldsymbol{x}(t)$ of $N \gg 1$ such particles in a time interval of duration t, and consider their projections $x(t)$ onto a fixed unit vector $\boldsymbol{e}$. We will get a distribution

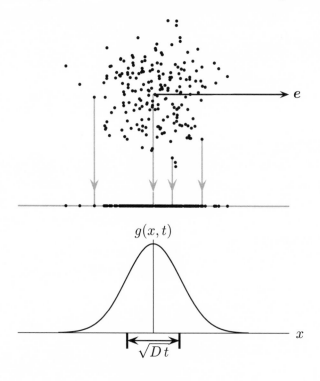

Figure 2.1.

$g(x, t)$ of these projections so that the fraction of $x(t)$ in $x_1 < x(t) < x_2$ is $\int_{x_1}^{x_2} g(x, t) \, dx$. By definition,

$$(2.1) \qquad \int_{-\infty}^{\infty} g(x, t) \, dx = 1.$$

Assume that the displacement process is *homogeneous*: The *same* $g(x, t)$ is obtained for displacements starting from different origins in space and time. Assume *isotropy*, which means the same $g(x, t)$ for different e's. In particular, the same $g(x, t)$ is obtained for e and for $-e$, so $g(x, t)$ is *even* in x. Because of the evenness, the *mean displacement* vanishes, i.e.,

$$(2.2) \qquad \langle x \rangle := \int_{-\infty}^{\infty} x g(x, t) \, dx = 0.$$

There is *independence*: Let x and x' be displacements of the same particle in two *nonoverlapping* time intervals of durations t and t'; the *joint distribution* of x and x' is $g(x, t)g(x', t')$.

The *mean square displacement* is observed to be proportional to time t:

$$(2.3) \qquad \langle x^2 \rangle := \int_{-\infty}^{\infty} x^2 g(x, t) \, dx = f(t) = 2Dt,$$

where D is a positive diffusion coefficient with units of $(\text{length})^2 \div \text{time}$. This property of $g(x, t)$ follows from the homogeneity, independence, and zero mean properties of Brownian motion. For $t, t' > 0$, divide the time interval $(0, t + t')$ into two pieces: $(0, t)$ and $(t, t + t')$. Let x and x' be the displacements in these two time intervals. The distributions of x and x' are $g(x, t)$ and $g(x', t')$, respectively. The net displacement $s := x + x'$ over time interval $(0, t + t')$ has distribution $g(s, t + t')$. The three g's mentioned here are the *same* by homogeneity. The mean square of s is determined from the statistics of x and x'. We have

$$(2.4) \qquad \langle s^2 \rangle = \langle (x + x')^2 \rangle = \langle x^2 \rangle + 2\langle xx' \rangle + \langle x'^2 \rangle.$$

Since the displacements x and x' happen in nonoverlapping time intervals, we have

$$\langle xx' \rangle = \int_{-\infty}^{\infty} \int_{-\infty}^{\infty} xx' g(x, t)g(x', t') \, dx \, dx' =$$
$$= \left(\int_{-\infty}^{\infty} x g(x, t) \, dx \right) \left(\int_{-\infty}^{\infty} x' g(x', t) \, dx' \right) = \langle x \rangle \langle x' \rangle = 0 \cdot 0 = 0.$$

As for the remaining terms in (2.4),

$$\langle x^2 \rangle = \int_{-\infty}^{\infty} x^2 g(x,t)\, dx := f(t),$$

$$\langle x'^2 \rangle = \int_{-\infty}^{\infty} x'^2 g(x',t')\, dx' = f(t'),$$

$$\langle s^2 \rangle = \int_{-\infty}^{\infty} s^2 g(s, t+t')\, ds = f(t+t'),$$

so (2.4) becomes

(2.5) $$f(t+t') = f(t) + f(t'), \quad \text{all } t, t' > 0.$$

It follows from (2.5) that $f(t)$ is indeed linear, as in (2.3).

Here is a heuristic treatment of higher moments $\langle x^n \rangle$, for $n > 2$: By evenness of $g(x,t)$ in x, odd moments (with n odd) are zero. In light of the mean square displacement property (2.3), it appears that the main support of $g(x,t)$ in x occupies an interval of linear size $\sqrt{Dt}$ about $x = 0$ (see Figure 2.1), so one expects even moments (with n even) to be proportional to $(Dt)^{n/2}$.

The diffusion PDE. Now suppose that at $t = 0$, the density of x's is a given nonnegative function $c(x,0)$. What is the density $c(x,t)$ of x's at time $t > 0$? Consider the particles at time $t + \tau$ which came from positions between x' and $x'+dx'$ at time t. The number of these particles is $c(x',t)dx'$, and their positions x at time $t + \tau$ have distribution

$$g(x - x', \tau),$$

as follows from the homogeneity of Brownian motion in space and time. Hence, their contribution to the density seen at position x and time $t + \tau$ is

(2.6) $$g(x - x', t)c(x', \tau)\, dx'.$$

Figure 2.2 displays this argument in visual form. Summing (2.6) over all dx' gives the total density of all particles at time $t + \tau$, irrespective of their positions at time t:

$$c(x, t+\tau) = \int_{-\infty}^{\infty} c(x',t)g(x - x', \tau)\, dx'.$$

Introduce $\zeta := x - x'$ as a new variable of integration in place of x'. Then

(2.7) $$c(x, t+\tau) = \int_{-\infty}^{\infty} c(x - \zeta, t)g(\zeta, \tau)\, d\zeta.$$

Now recall that the main support of $g(\zeta, \tau)$ in ζ resides in an interval of size $\sqrt{D\tau}$ about $\zeta = 0$, so it makes sense to Taylor expand $c(x - \zeta, \tau)$ in powers

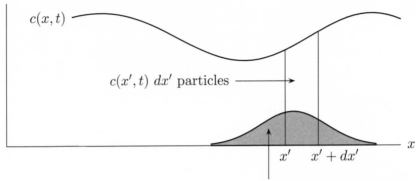

$c(x,t)$

$c(x', t)\, dx'$ particles

x' $x' + dx'$

x

Their density at time $t + \tau$ is
$g(x - x', \tau)\, c(x', t)\, dx'$.

Figure 2.2.

of ζ. So formally,

$$(2.8) \quad c(x, t + \tau) = \int_{-\infty}^{\infty} \left\{ c(x,t) - \zeta c_x(x,t) + \frac{\zeta^2}{2}\, c_{xx}(x,t) - \right.$$

$$\left. \frac{\zeta^3}{6}\, c_{xxx}(x,t) + \frac{\zeta^4}{24}\, c_{xxxx}(x,t) + \cdots \right\} g(\zeta, \tau)\, d\zeta$$

$$= c(x,t)\langle 1 \rangle - c_x(x,t)\langle \zeta \rangle + \frac{1}{2}\, c_{xx}(x,t)\langle \zeta^2 \rangle -$$

$$\frac{1}{6}\, c_{xxx}(x,t)\langle \zeta^3 \rangle + \frac{1}{24}\, c_{xxxx}(x,t)\langle \zeta^4 \rangle - \cdots .$$

In (2.8), $\langle \zeta^n \rangle := \int_{-\infty}^{\infty} \zeta^n g(\zeta, \tau)\, d\zeta$ are the moments of $g(\zeta, \tau)$. We have $\langle 1 \rangle = 1$ by (2.1) and $\langle \zeta^2 \rangle = 2D\tau$ by (2.3). Odd moments vanish by evenness of $g(\zeta, \tau)$ in ζ, and the higher even moments $\langle \zeta^n \rangle$, n even, are of size $(D\tau)^{n/2}$. Hence, (2.8) reduces to

$$(2.9) \quad c(x, t + \tau) - c(x, t) = D\tau c_{xx}(x,t) + O(\tau^2).$$

Here, $O(\tau^2)$ means that all the other terms not explicitly mentioned in (2.9) make up a sum which is bounded in absolute value by a constant times τ^2. Taylor expanding the left-hand side of (2.9) in powers of τ results in

$$c_t(x,t) - Dc_{xx}(x,t) = O(\tau),$$

and the limit $\tau \to 0$ gives the *diffusion PDE*

$$(2.10) \quad c_t = Dc_{xx},$$

which holds for all x and for $t > 0$. The required solution $c(x,t)$ matches the given initial condition $c(x,0)$ at $t = 0$.

Problems 2.1–2.5 present the lattice random walk model of diffusion.

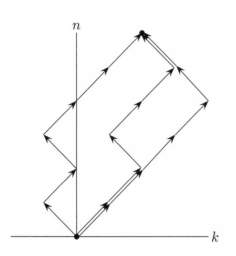

Figure 2.3.

Problem 2.1. In each "time step" a particle makes a displacement of fixed length along a line in the positive or negative direction. Let the time step be the unit of time, and let the displacement in one time step be the unit of length, so that the duration and displacement of each step are both one. Figure 2.3 shows trajectories of particles in spacetime starting from $(0,0)$ and ending at (k,n). For any k with $|k| < n$, there are multiple trajectories with the same endpoint, and we begin by *counting* them. Let $N(k,n)$ be the number of distinct spacetime paths from $(0,0)$ to (k,n). Show that $N(k,n)$ satisfies a "discrete initial value problem"

$$N(k, n+1) = N(k-1, n) + N(k+1, n),$$

$$N(k,1) = \begin{bmatrix} 1 & : & k = 1 \text{ or } -1, \\ 0 & : & \text{otherwise.} \end{bmatrix}$$

Show from this initial value problem that the number of all paths of n steps starting from $(0,0)$ and ending up at any position k between $-n$ and n is 2^n. Hence,

$$f(k,n) := \frac{1}{2^n} N(k,n)$$

is the fraction of walks of n steps that end up at position k. Reformulate the initial value problem for $N(k,n)$ in terms of $f(k,n)$.

Solution. If the particle is at position k after $n+1$ steps, it was at position $k-1$ or $k+1$ after n steps. Hence,

(2.1-1) $$N(k, n+1) = N(k-1, n) + N(k+1, n).$$

Starting from $k=0$ at $n=0$, there is one path to $k=1$ at $n=1$, and one path to $k=-1$. No *other* positions k are possible at $n=1$. Hence,

$$N(k,1) = \begin{bmatrix} 1 & : & k = 1 \text{ or } -1, \\ 0 & : & \text{otherwise.} \end{bmatrix}$$

Summation of (2.1-1) with respect to k gives

$$\sum_k N(k, n+1) = \sum_k N(k-1, n) + \sum_k N(k+1, n) = 2 \sum_k N(k, n),$$

and hence,

$$\sum_k N(k, n) = 2^{n-1} \sum_k N(k, 1) = 2^{n-1} \cdot 2 = 2^n.$$

The initial value problem for $f(k, n) := 2^{-n} N(k, n)$ is

$$f(k, n+1) \;=\; \frac{1}{2} f(k-1, n) + \frac{1}{2} f(k+1, n), \quad n \geq 1,$$

$$f(k, 1) \;=\; \left[\begin{array}{lll} \frac{1}{2} & : & k = 1 \text{ or } -1, \\ 0 & : & \quad\text{otherwise.} \end{array} \right.$$

Problem 2.2.   Define $\langle k^2 \rangle(n) := \sum_k k^2 f(k, n)$ (the discrete mean square displacement). Show from the initial value problem for $f(k, n)$ that

$$\langle k^2 \rangle(n) = n.$$

Solution.

$$\langle k^2 \rangle(n+1) \;=\; \sum_k k^2 f(k, n+1)$$

$$=\; \frac{1}{2} \sum_k k^2 f(k-1, n) + \frac{1}{2} \sum_k k^2 f(k+1, n)$$

$$=\; \sum_k \frac{(k+1)^2 + (k-1)^2}{2} f(k, n)$$

$$=\; \sum_k (k^2 + 1) f(k, n)$$

$$=\; \langle k^2 \rangle(n) + 1, \quad n \geq 1.$$

The last equality uses $\sum_k f(k, n) = 1$. Hence,

$$\langle k^2 \rangle(n) = n - 1 + \langle k^2 \rangle(1) = n - 1 + 1 = n.$$

Problem 2.3.   We derive a *continuum limit* of the discrete initial value problem, valid for large k and n. The continuum representation of f is sought in the form

(2.3-1) $$f(k,n) = \varepsilon^p g(x := \varepsilon k, t := \varepsilon^q n, \varepsilon).$$

Here, g is a smooth function of its arguments and ε is a "gauge parameter". The exponents p and q are to be determined so that

$$\varepsilon^{-p} f(k,n) \to g^0(x,t) := g(x,t,\varepsilon = 0)$$

in the limit $\varepsilon \to 0$ with $x = \varepsilon k$ and $t = \varepsilon^q n$ fixed. Note: From the original discrete initial value problem, we see that the support of f in k is even k's for even n's and odd k's for odd n's, so (2.3-1) really applies on the support of $f(k,n)$. We want to determine p and g.

a) Determine the exponent p such that

$$\int_{-\infty}^{\infty} g(x,t,\varepsilon)dx \to 1$$

as $\varepsilon \to 0$. In this case, $g^0(x,t) := g(x,t,\varepsilon = 0)$ satisfies the integral condition

$$\int_{-\infty}^{\infty} g^0(x,t)dx = 1.$$

b) Determine the exponent q such that $\langle k^2 \rangle = n$ translates into $\langle x^2 \rangle = \langle t \rangle$.

Solution.

a) We have

$$1 = \sum_k f(k,n) = \varepsilon^p \sum_k g(\varepsilon k, t, \varepsilon) = \varepsilon^{p-1} \sum_k g(\varepsilon k, t, \varepsilon)\varepsilon.$$

In the right-hand side, $\sum_k g(\varepsilon k, t, \varepsilon)\varepsilon$ is a Riemann sum for $\int_{-\infty}^{\infty} g(x,t,\varepsilon)dx$, and it is evident that $p = 1$.

b) Rewrite $\langle k^2 \rangle(n) = \sum_k k^2 f(k,n) = n$ as

$$\sum_k (\varepsilon k)^2 g(\varepsilon k, t, \varepsilon)\varepsilon = \varepsilon^{2-q} t.$$

The Riemann sum for $\langle x^2 \rangle = \int_{-\infty}^{\infty} x^2 g(x,t,\varepsilon)dx$ appears on the left-hand side. Hence, we require $q = 2$. In summary, the continuum representation of f is

$$f(k,n) = \varepsilon g(x := \varepsilon k, t := \varepsilon^2 n, \varepsilon).$$

Problem 2.4. Now substitute (2.3-1) (with your values of p and q) into the recursion equation for $f(k, n)$ in Problem 2.1. Take the limit $\varepsilon \to 0$ in order to derive the PDE for $g^0(x, t)$. This PDE is subject to the integral condition $\int_{\infty}^{\infty} g^0(x, t) dx = 1$ and initial condition $g^0(x, t) \to 0$ as $t \to 0$ with $x \neq 0$ fixed. The initial condition means that "there are very few long paths in short time". The solution of this initial value problem is called the *diffusion kernel*.

Solution. We introduce substitutions

$$x = \varepsilon k, \qquad t = \varepsilon^2 n, \qquad f(k, n) = \varepsilon g(x, t, \varepsilon)$$

into the recursion relation

$$f(k, n+1) = \frac{1}{2} f(k-1, n) + \frac{1}{2} f(k+1, n)$$

to obtain

(2.4-1) $g(x, t + \varepsilon^2, \varepsilon) = \frac{1}{2} g(x + \varepsilon, t, \varepsilon) + \frac{1}{2} g(x - \varepsilon, t, \varepsilon).$

By Taylor expansion, we have

$$
\begin{aligned}
g(x, t + \varepsilon^2, \varepsilon) &= g(x, t, \varepsilon) + \varepsilon^2 g_t(x, t, \varepsilon) + O(\varepsilon^4), \\
\frac{1}{2} g(x + \varepsilon, t, \varepsilon) + \frac{1}{2} g(x - \varepsilon, t, \varepsilon) &= g(x, t, \varepsilon) + \frac{\varepsilon^2}{2} g_{xx}(x, t, \varepsilon) + O(\varepsilon^4),
\end{aligned}
$$

so (2.4-1) becomes

$$g_t(x, t, \varepsilon) = \frac{1}{2} g_{xx}(x, t, \varepsilon) + O(\varepsilon^2),$$

and in the limit $\varepsilon \to 0$,

$$g_t^0(x, t) = \frac{1}{2} g_{xx}^0(x, t).$$

Problem 2.5. If the length of each space step is ℓ and the duration of each time step is τ, what is the actual dimensional diffusion coefficient associated with the lattice random walk?

Solution. In $\langle k^2 \rangle = n$, recall that k is displacement in units of ℓ, and n is time in units of τ. Hence, dimensional displacement and time are given by $x = k\ell$ and $t = n\tau$, and $\langle k^2 \rangle = n$ translates into $\langle x^2 \rangle = \frac{\ell^2}{\tau} t$. Comparing with $\langle x^2 \rangle = 2Dt$, we see that $D = \frac{\ell^2}{2\tau}$.

Diffusion in $\mathbb{R}^n$. The $\mathbb{R}^2$ case is sufficient to make the point. Let e_1, e_2 be cartesian unit vectors of $\mathbb{R}^2$, and let $x_1(t)$ and $x_2(t)$ represent the e_1 and e_2 displacements of Brownian particles in time t. We assume that e_1 and e_2 displacements are independent, so that the joint distribution of $x_1(t)$ and $x_2(t)$ is $g(x_1, t)g(x_2, t)$; that is, the fraction of $(x_1(t), x_2(t))$ in region R of $\mathbb{R}^2$ is $\int_R g(x_1, t)g(x_2, t)\, dx_1\, dx_2$. Now let $c(x_1, x_2, t)$ be the density of Brownian particles at point (x_1, x_2) at time t. The $\mathbb{R}^2$ version of the integral equation (2.7) is

$$(2.11) \quad c(x_1, x_2, t + \tau) = \int_{\mathbb{R}^2} c(x_1 - \zeta_1, x_2 - \zeta_2, t)g(\zeta_1, t)g(\zeta_2, t)\, d\zeta_1\, d\zeta_2.$$

Now write the Taylor series of $c(x_1 - \zeta_1, x_2 - \zeta_2, t)$ in powers of ζ_1, ζ_2 and substitute it into (2.11). We get the $\mathbb{R}^2$ counterpart of (2.8),

$$(2.12) \quad c(x_1, x_2, t + \tau) = c(x_1, x_2, t) - \partial_1 c(x_1, x_2, t)\langle \zeta_1 \rangle - \partial_2 c(x_1, x_2, t)\langle \zeta_2 \rangle$$
$$+ \frac{1}{2}\partial_{11} c(x_1, x_2, t)\langle \zeta_1^2 \rangle + \partial_{12} c(x_1, x_2, t)\langle \zeta_1 \zeta_2 \rangle +$$
$$+ \frac{1}{2}\partial_{22} c(x_1, x_2, t)\langle \zeta_2^2 \rangle + \cdots.$$

As before, $\langle \zeta_1 \rangle = \langle \zeta_2 \rangle = 0$ and $\langle \zeta_1^2 \rangle = \langle \zeta_2^2 \rangle = 2D\tau$, and by the independence of e_1 and e_2 displacements, $\langle \zeta_1 \zeta_2 \rangle = \langle \zeta_1 \rangle \langle \zeta_2 \rangle = 0$. Hence, (2.12) reduces to

$$c(x_1, x_2, t + \tau) - c(x_1, x_2, t) = D\tau(\partial_{11}c + \partial_{22}c)(x_1, x_2, t) + O(\tau^2).$$

Now divide by τ and take the limit as $\tau \to 0$ to get the diffusion equation on $\mathbb{R}^2$,

$$(2.13) \qquad\qquad \partial_t c = D(\partial_{11}c + \partial_{22}c).$$

In general, the diffusion equation on $\mathbb{R}^n$ is

$$(2.14) \qquad\qquad \partial_t c = D(\partial_{11}c + \cdots + \partial_{nn}c) := D\Delta c.$$

Fick's law. Let R be a fixed region of $\mathbb{R}^n$. Integrate (2.14) over R and use the divergence theorem to get *Fick's law*:

$$(2.15) \qquad \frac{d}{dt}\int_R c\, d\boldsymbol{x} = \int_R D\Delta c\, d\boldsymbol{x} = -\int_{\partial R} (-D\boldsymbol{\nabla} c) \cdot \boldsymbol{n}\, da.$$

The vector field

$$(2.16) \qquad\qquad \boldsymbol{f} := -D\boldsymbol{\nabla} c$$

is called the *diffusive flux*. Equation (2.15) says that Brownian particles enter R by crossing the boundary ∂R at a rate per unit area:

$$-\boldsymbol{f} \cdot \boldsymbol{n} = D\boldsymbol{\nabla} c \cdot \boldsymbol{n} = Dc_n.$$

Here,

$$(2.17) \qquad\qquad c_n := \boldsymbol{\nabla} c \cdot \boldsymbol{n}$$

is the *normal derivative* of c on ∂R. Figure 2.4 illustrates Fick's law (2.15) in the $\mathbb{R}^2$ case.

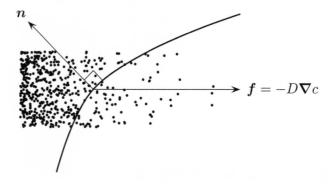

Figure 2.4.

Problem 2.6 (Lyapunov functional for a diffusion process).   Particles in the exterior $R := \mathbb{R}^3 - \bar{R}$ of a bounded region $\bar{R}$ diffuse with diffusivity D. For times $t < 0$, the boundary ∂R is an impenetrable wall, and there is uniform concentration $c \equiv c_0$ in R. For $t > 0$, ∂R becomes a source of particles: $s(\boldsymbol{x})$ defined for $\boldsymbol{x}$ on ∂R is the rate per unit area at which particles are "launched" from ∂R into R. Formulate the boundary value problem for the particle concentration in R, and show that the Lyapunov functional

$$(2.6\text{-}1) \qquad E[c] := -\int_{\partial R} sc\, da + \frac{D}{2} \int_R |\boldsymbol{\nabla} c|^2 d\boldsymbol{x}$$

is decreasing in time wherever c_t is not identically zero.

Solution. The particle concentration $c(\boldsymbol{x}, t)$ satisfies the diffusion PDE

$$c_t = D\Delta c$$

in R. The rate of influx of particles from ∂R is expressed by the boundary condition

$$Dc_n = -s.$$

Here, the unit normal is taken to be *outward* from $\bar{R}$, hence the minus sign. Finally there is the "boundary condition at ∞",

$$c \to c_0 \qquad \text{as} \qquad |\boldsymbol{x}| \to \infty,$$

because at any finite time, the fraction of introduced particles which diffuse distances much greater than $\sqrt{Dt}$ is negligible. Time differentiation of (2.6-1) gives

(2.6-2) $$\dot{E} = -\int_{\partial R} sc_t \, da + D \int_R \boldsymbol{\nabla} c \cdot \boldsymbol{\nabla} c_t \, d\boldsymbol{x}.$$

Integrating the "product rule" identity

$$\boldsymbol{\nabla} \cdot (c_t \boldsymbol{\nabla} c) = c_t \Delta c + \boldsymbol{\nabla} c_t \cdot \boldsymbol{\nabla} c$$

over R and using the divergence theorem gives

(2.6-3) $$-\int_{\partial R} c_t c_n \, da = \int_R \Delta c \, c_t \, d\boldsymbol{x} + \int_R \boldsymbol{\nabla} c \cdot \boldsymbol{\nabla} c_t \, d\boldsymbol{x}.$$

The minus sign in front of the surface integral is due to $\boldsymbol{n}$ pointing *into* R. Using (2.6-3) to substitute for $\int_R \boldsymbol{\nabla} c \cdot \boldsymbol{\nabla} c_t \, d\boldsymbol{x}$ in (2.6-2) gives

$$\dot{E} = -\int_{\partial R} (s + Dc_n) c_t \, da - D \int_R \Delta c \, c_t \, d\boldsymbol{x}.$$

The surface integral vanishes by the boundary condition on ∂R, and putting $D\Delta c = c_t$ gives

$$\dot{E} = -\int_R c_t^2 \, d\boldsymbol{x},$$

which is negative whenever c_t is not identically zero.

Problem 2.7 (The Maxwell flux).   The result of Problem 2.6 suggests that $c(\boldsymbol{x}, t)$ in R relaxes to the minimum of the Lyapunov functional $E[c]$ as $t \to \infty$, and that this minimum is a solution $c(\boldsymbol{x})$ of the time-independent boundary value problem. Solve the time-independent boundary value problem for the case of a sphere of radius a, with s being a uniform constant on $r = a$. Show that the net rate F at which particles are introduced into $r > a$ is proportional to $c(a) - c_0$ and a.

Solution. $c = c(r)$ is a spherically symmetric harmonic function which asymptotes to c_0 as $r \to \infty$. Hence,

$$c(r) = c_0 + (c(a) - c_0)\frac{a}{r}.$$

The total outflux of particles from the sphere is

$$F - 4\pi a^2 Dc_r(a) = 4\pi a D(c(a) - c_0).$$

The famous English physicist Maxwell encountered and solved this boundary value problem while studying precipitation of solids from liquid solutions.

Problem 2.8 (Steady state in two dimensions?).   We consider the same process as in Problem 2.6, but in $\mathbb{R}^2$ instead of $\mathbb{R}^3$. The Brownian particles are in the portion of the plane exterior to a circle of radius a. Does $c(r,t)$ relax to a time-independent concentration $c(r)$ as $t \to \infty$?

Solution. No. The most general circularly symmetric harmonic function in $r > a$ which satisfies the boundary condition $Dc_r(a) = -s$ is $c(r) = -\frac{sa}{D}\log\frac{r}{a} + \alpha$. For no value of the arbitrary constant α can we satisfy $c(\infty) = c_0$. Later, Problem 2.15 will address the (nontrivial) long-time dependence of $c(r,t)$.

Melting interfaces. Liquid and solid phases of a substance occupy complementary regions $R(t)$ and $\bar{R}(t)$ of $\mathbb{R}^3$. We present the simplest model for the motion of the solid-liquid interface as dictated by the heat transfer between solid and liquid. The solid phase is at the melting temperature, and $h(\boldsymbol{x},t)$ in $R(t)$ denotes the heat content per unit volume of the liquid phase, relative to liquid at the melting temperature. The heat content h satisfies the diffusion PDE

$$(2.18) \qquad\qquad h_t = D\Delta h$$

in $R(t)$, where D is the thermal diffusivity. The liquid at the interface assumes the melting temperature (same as the solid), so we have the boundary condition

$$(2.19) \qquad\qquad h = 0$$

on $\partial R(t)$. When the solid melts, there is generally a volume change which we will neglect, and the melting of a volume V of solid requires an input of heat HV, where the constant $H > 0$ is called the latent heat of fusion. Hence, conservation of thermal energy is expressed by

$$(2.20) \qquad\qquad \frac{d}{dt}\int_{R(t)} h(\boldsymbol{x},t)d\boldsymbol{x} = -H\dot{V}(t),$$

where $V(t)$ is the volume of liquid in $R(t)$. By the boundary condition $h = 0$ on ∂R and the diffusion PDE (2.18) in R, the left-hand side of (2.20) is

$$\int_R h_t\,d\boldsymbol{x} = D\int_R \Delta h\,d\boldsymbol{x} = D\int_{\partial R} h_n\,da,$$

and the right-hand side is

$$-H \int_{\partial R} v \, da,$$

where v is the normal velocity of ∂R. Hence, (2.20) is rewritten as

$$\int_{\partial R} (Dh_n + Hv) da = 0,$$

and we surmise from the energy conservation (2.20) the heat transfer boundary condition

(2.21) $Dh_n = -Hv$

on ∂R. Notice that if we reverse the orientation of the unit normal $\boldsymbol{n}$, then the normal derivative h_n and the normal velocity v both change sign, and the boundary condition (2.21) is invariant. The diffusion PDE (2.18) together with the two boundary condition on ∂R is a *free boundary problem* for the evolutions of $h(\boldsymbol{x}, t)$ in $R(t)$ *and* the solid-liquid interface $\partial R(t)$.

We mention one oversimplification that restricts the usefulness of the simplified free boundary problem to melting interfaces: The interface temperature is generally lower than the melting temperature by an amount proportional to the local interface mean curvature. This role of curvature should be small if the characteristic length of interface geometry is sufficiently large. "Melting" interfaces are generally stable against the growth of small-scale roughness, but "freezing" interfaces are unstable. Hence, the free boundary problem presented here is a simplified model of melting interfaces.

Let us consider a finite region $\bar{R}(t)$ of solid surrounded by unbounded liquid in $R(t) = \mathbb{R}^3 - \bar{R}(t)$ with $h \to h_\infty > 0$ as $|\boldsymbol{x}| \to \infty$. The "excess" heat content h_∞ per unit volume drives the melting of the solid. The free boundary problem simplifies in the *quasistatic* limit

$$\varepsilon := \frac{h_\infty}{H} \to 0.$$

The reduction begins with a nondimensionalization of the equations based on units in the following *scaling table*:

Variable	h	$\boldsymbol{x}$	v	t
Unit	h_∞	a	$\frac{D}{a}\frac{h_\infty}{H}$	$\frac{a^2}{D}\frac{H}{h_\infty}$

Here, a is a characteristic length of the solid region $\bar{R}$. The unit of normal velocity v comes from an order-of-magnitude balance between the left-hand side and right-hand side of the heat transfer boundary condition (2.21). Assuming that the characteristic length associated with $h(\boldsymbol{x}, t)$ is a (comparable to the dimensions of the solid), the left-hand side has order of magnitude $\frac{Dh_\infty}{a}$ and the balance with Hv on the right-hand side gives the unit of v.

There is an implicit assumption that the time dependence of $h(\boldsymbol{x},t)$ is dominated by the motion of the interface: The unit $\frac{a^2}{D}\frac{H}{h_\infty}$ of time is a divided by the unit of v. The "relaxation time" for diffusion in a "zone" of thickness a about the solid is $\frac{a^2}{D}$, much shorter. The dimensionless versions of the diffusion PDE (2.18) and the heat transfer boundary condition (2.21) are

$$(2.22) \qquad \Delta h = \varepsilon h_t$$

in $R(t)$, and

$$(2.23) \qquad h_n = v$$

on $\partial R(t)$. The zero boundary condition (2.19) is of course invariant under scaling, and the boundary condition $h \to h_\infty$ as $|\boldsymbol{x}| \to \infty$ translates into $h \to 1$ as $|\boldsymbol{x}| \to \infty$. The scaling of the equations is an elementary process, but a little demonstration of technique is given. For instance, write the *dimensional* diffusion PDE as

$$(2.24) \qquad D\Delta h = (\partial_t)h.$$

We replace each dimensional quantity or operator by the appropriate scaling unit times its nondimensional counterpart. Hence, h is replaced by $h_\infty h$, the Laplacian Δ by $\frac{1}{a^2}\Delta$, and ∂_t by $\frac{D}{a^2}\frac{h_\infty}{H}\partial_t$. The result of these substitutions in (2.24) is (2.22). The quasistatic limit is obtained by setting $\varepsilon = 0$ in the dimensionless equations. This means that the diffusion PDE (2.22) is replaced by Laplace's equation. In solving little examples, it is sometimes convenient to rewrite the limit equations in original dimensional units. Here, we see that the quasistatic limit amounts to simply striking out the time derivative in the diffusion PDE (2.18).

We finish with a simple example: $\bar{R}$ is a sphere of radius $a(t)$. At each time t, $h = h(r,t)$ is the spherically symmetric harmonic function which has $h(a,t) = 0$, $h(\infty,t) = h_\infty$; that is,

$$h(r,t) = h_\infty\left(1 - \frac{a}{r}\right)$$

in $r > a$. The heat transfer boundary condition (2.21) now gives

$$Dh_r(a,t) = \frac{Dh_\infty}{a} = -H\dot{a},$$

or

$$(2.25) \qquad \dot{a} = -\frac{h_\infty}{H}\frac{D}{a}.$$

From this ODE, it readily follows that a solid sphere of initial radius a melts completely in time

$$(2.26) \qquad t = \frac{1}{2}\frac{H}{h_\infty}\frac{a^2}{D}.$$

Sometimes, a formal reduction like this quasistatic limit is deceiving. The same scaling reduction can be done formally in the one- or two-dimensional case. For instance, consider the melting of a solid *slab* of thickness $2a$. Let x be distance from the center of the slab. Harmonic functions that depend only on x are *linear* in x, so we cannot satisfy both boundary conditions $h = 0$ on $x = a$ and $h \to h_\infty > 0$ as $x \to \infty$. The time derivative in the diffusion PDE *must* be essential. If the slab is placed into the liquid at time $t = 0$, we expect that at times $t > 0$, the influence of heat transfer to the slab will be felt at distances $\sqrt{Dt}$ from the center. An order-of-magnitude balance in the heat transfer boundary condition (2.21), based on the length scale $\sqrt{Dt}$, gives an estimate for the slab's rate of shrinkage,

$$\dot{a} \approx -\frac{h_\infty}{H}\sqrt{\frac{D}{t}}.$$

Specifically, we replaced h_n by $\frac{h_\infty}{\sqrt{Dt}}$. The time t to melt a slab of initial thickness $2a$ is estimated to be

$$(2.27) \qquad\qquad t \approx \left(\frac{H}{h_\infty}\right)^2 \frac{a^2}{D}.$$

In the limit $\frac{h_\infty}{H} \to 0$, this result is correct modulo a numerical prefactor, as we shall see from a later, rigorous solution of this slab problem (Problem 2.14). The melting of a cylindrical solid of radius a also confounds the quasistatic limit. Recall that circularly symmetric harmonic functions on $\mathbb{R}^2$ diverge like $\log r$ as $r \to \infty$. So, the time derivative in the diffusion PDE is essential in the two-dimensional case as well.

.

Problems 2.9–2.12 analyze a nonlinear free boundary problem.

Problem 2.9. Particles diffuse in $\mathbb{R}^3$, with a small diffusion coefficient for high concentrations and a large diffusion coefficient for low concentrations, that is,

$$D = \left[\begin{array}{ll} D_1 & \text{for } c(\boldsymbol{x}, t) > c_0, \\ D_2 > D_1 & \text{for } c(\boldsymbol{x}, t) < c_0. \end{array} \right.$$

Let $R(t)$ be a high-concentration region, with $c = c_1(\boldsymbol{x}, t) > c_0$, surrounded by low concentration $c = c_2(\boldsymbol{x}, t) < c_0$ in $\bar{R}(t) = \mathbb{R}^3 - R(t)$. Formulate the free boundary problem which governs the evolutions of c_1 in R, c_2 in $\bar{R}$, and the interface ∂R.

Solution. We have a diffusion PDE for c_1 in R $(D = D_1)$ and for c_2 in $\bar{R}$ $(D = D_2)$. A diffusion process cannot sustain a jump discontinuity in concentration, so

$$(2.9\text{-}1) \qquad\qquad c_1 = c_0 = c_2 \quad \text{on } \partial R.$$

In addition, the normal components of diffusive fluxes are continuous, so

$$(2.9\text{-}2) \qquad\qquad D_1 \partial_n c_1 = D_2 \partial_n c_2 \quad \text{on } \partial R.$$

Here is the calculation behind (2.9-2): By conservation of particles,

$$(2.9\text{-}3) \qquad\qquad \frac{d}{dt} \int_R c_1 \, d\boldsymbol{x} + \frac{d}{dt} \int_{\bar{R}} c_2 \, d\boldsymbol{x} = 0.$$

We have

$$(2.9\text{-}4) \qquad \begin{aligned} \frac{d}{dt} \int_R c_1 \, d\boldsymbol{x} &= \int_R \partial_t c_1 \, d\boldsymbol{x} + c_0 \int_{\partial R} v \, da \\ &= \int_{\partial R} \{D_1 \partial_n c + c_0 v\} \, da. \end{aligned}$$

Here, the normal derivative $\partial_n c_1$ and normal velocity v assume outward orientation of $\boldsymbol{n}$ with respect to R. Similarly,

$$(2.9\text{-}5) \qquad\qquad \frac{d}{dt} \int_{\bar{R}} c_2 \, d\boldsymbol{x} = -\int_{\partial R} \{D_2 \partial_n c_2 + c_0 v\} \, d\boldsymbol{x}.$$

There is a minus sign because the unit normal of $\partial \bar{R}$ is $-\boldsymbol{n}$. Substituting (2.9-4) and (2.9-5) into (2.9-3), we get

$$\int_{\partial R} \{D_1 \partial_n c_1 - D_2 \partial_n c_2\} \, da = 0,$$

from which we surmise (2.9-2). The free boundary problem consists of the diffusion PDEs for c_1 and c_2 and the boundary conditions (2.9-1) and (2.9-2).

Problem 2.10. We examine a limit with $\varepsilon := \frac{D_1}{D_2} \to 0$. The magnitudes of c_1 in R and c_2 in $\bar{R}$ are $\frac{c_0}{\varepsilon}$ and c_0. Let a be a characteristic length of R, and assume that the characteristic lengths of $\boldsymbol{x}$-variations of c_1 and c_2 are both a. What is the characteristic time t_c of interface motion? In other words, how long does it take for the interface to move a distance comparable to a? Nondimensionalize the free boundary problem using the following units:

Variable	$\boldsymbol{x}$	t	c_1	c_2
Unit	a	t_c	$\frac{c_0}{\varepsilon}$	c_0

Take the $\varepsilon \to 0$ limit of the dimensionless equations.

Solution. The rate of loss of particles from R has magnitude $\left(\frac{1}{t_c}\right)\left(\frac{c_0}{\varepsilon}a^3\right)$. These lost particles appear in $\bar{R}$. The flux $D_2\partial_n c_2$ into $\bar{R}$ across ∂R has magnitude $D_2\frac{c_0}{a}$, and the surface integral of this flux over ∂R has magnitude $\left(D_2\frac{c_0}{a}\right)(a^2)$. The balance $\left(\frac{1}{t_c}\right)\left(\frac{c_0}{\varepsilon}a^3\right) = \left(D_2\frac{c_0}{a}\right)(a^2)$ gives $t_c = \frac{1}{\varepsilon}\frac{a^2}{D_2} = \frac{a^2}{D_1}$. The derivation of the dimensionless diffusion PDE for c_1 in R starts with

$$\left(\frac{1}{t_c}\partial_t\right)\left(\frac{c_0}{\varepsilon}c_1\right) = D_1\left(\frac{1}{a^2}\Delta\right)\left(\frac{c_0}{\varepsilon}c_1\right).$$

Given that $t_c = \frac{a^2}{D_1}$, this reduces to

(2.10-1) $\qquad\qquad\qquad \partial_t c_1 = \Delta c_1 \quad$ in R.

Similarly,

(2.10-2) $\qquad\qquad\qquad \varepsilon \partial_t c_2 = \Delta c_2 \quad$ in $\bar{R}$.

We expect (2.10-2) since the diffusion coefficient in $\bar{R}$ is $\frac{1}{\varepsilon}$ times the diffusion coefficient in R. The dimensionless version of the boundary condition $c_1 = c_0 = c_2$ on ∂R is $\left(\frac{c_0}{\varepsilon}\right) c_1 = c_0 = (c_0)c_2$, or

(2.10-3) $\qquad\qquad\qquad c_1 = \varepsilon, \quad c_2 = 1 \quad$ on ∂R.

The dimensionless version of $D_1\partial_n c_1 = D_2\partial_n c_2$ on ∂R is $D_1\left(\frac{1}{a}\partial_n\right)\left(\frac{c_0}{\varepsilon}c_1\right) = D_2\left(\frac{1}{a}\partial_n\right)(c_0 c_2)$ or, using $D_1 = \varepsilon D_2$,

(2.10-4) $\qquad\qquad\qquad \partial_n c_1 = \partial_n c_2 \quad$ on ∂R.

The $\varepsilon \to 0$ limit equations are:

(2.10-5) $\qquad \partial_t c_1 = \Delta c_1 \quad$ in R, $\qquad c_1 = 0 \quad$ on ∂R,

(2.10-6) $\qquad \Delta c_2 = 0 \quad$ in $\bar{R}$, $\qquad c_2 = 1 \quad$ on ∂R,

(2.10-7) $\qquad \partial_n c_1 = \partial_n c_2 \quad$ on ∂R.

Problem 2.11. We examine solutions of the $\varepsilon \to 0$ limit equations in Problem 2.10 assuming that R is a sphere of (dimensionless) radius $a(t)$ and that c_1 and c_2 are spherically symmetric, with $c_2 \to 0$ as $r \to \infty$. Derive a reduced free boundary problem which governs the evolution of $a(t)$ and of $u(r,t) := rc_1(r,t)$ in $0 < r < a$.

Solution. The spherically symmetric diffusion PDE for c_1 can be written as

$$\partial_t(rc_1) = r(\partial_{rr}c_1 + \frac{2}{r}\partial_r c_1) = \partial_{rr}(rc_1).$$

Setting $u = rc_1$, we get

(2.11-1) $$\partial_t u = \partial_{rr}u \quad \text{in } 0 < r < a(t).$$

Assuming c_1 is bounded as $r \to 0$, we get a zero boundary condition on u at $r = 0$, and $c_1 = 0$ on $r = a$ gives a zero boundary condition on u at $r = a$ as well:

(2.11-2) $$u(0,t) = u(a,t) = 0.$$

There is a derivative boundary condition at $r = a$: We have $(\partial_r u)(a,t) = c_1(a,t) + a(\partial_r c_1)(a,t) = 0 + a(\partial_r c_2)(a,t)$. Now c_2 in $r > a$ is the spherically symmetric harmonic function with values equal to 1 at $r = a$ and 0 at $r = \infty$. Hence, $c_2 = \frac{a}{r}$ in $r > a$ and $\partial_r c_2 = -\frac{1}{a}$ at $r = a$, so finally,

(2.11-3) $$(\partial_r u)(a,t) = -1.$$

Our reduced free boundary problem consists of equations (2.11-1)–(2.11-3).

Problem 2.12. We seek $u(r,t)$ in "profile" form, $u = b(t)U(\zeta := \frac{r}{a(t)})$. Show that $a\,\dot{a} = -\lambda$, where λ is a *positive* constant. How is λ determined? Show that $R : 0 \leq r < a$ collapses to a point in finite time, and that particles are lost at a rate proportional to $a(t)$.

Solution. First, the boundary condition $(\partial_r u)(a,t) = -1$ implies $\frac{b(t)}{a(t)}U'(1) = -1$. Hence, $b(t) = a(t)$ modulo a multiplicative constant. Without loss of generality, we take the constant to be 1, so

(2.12-1) $$U'(1) = -1.$$

Substituting $u = a(t)U(\zeta = \frac{r}{a(t)})$ into the diffusion PDE $\partial_t u = \partial_{rr}u$ gives

$$\dot{a}(U - \zeta U') = \frac{1}{a}U''$$

or

$$U'' - (a\,\dot{a})(U - \zeta U') = 0.$$

The usual separation of variables argument implies that

(2.12-2) $$a\,\dot{a} = -\lambda,$$

where λ is a constant. Hence, we have an eigenvalue problem for λ and $U(\zeta)$,

(2.12-3)
$$U'' + \lambda(U - \zeta U') = 0 \quad \text{in } 0 < \zeta < 1,$$
$$U(0) = U(1) = 0.$$

Notice that the eigenfunction $U(\zeta)$ is normalized by $U'(1) = -1$, as in (2.12-1). Since $c_1 = \frac{a}{r}U(\frac{r}{a})$ is to be positive in $0 < r < a$, we require the "ground state" eigenfunction $U_1(\zeta)$ with the smallest eigenvalue λ_1. We show that λ_1 is positive. Multiply the ODE in (2.12-3) by U and integrate over $0 < \zeta < 1$. After the usual integration by parts and use of zero boundary conditions, we find that

$$\frac{3}{2}\lambda_1 \int_0^1 U^2 \, d\zeta = \int_0^1 U'^2 \, d\zeta,$$

so $\lambda_1 > 0$.

We go back to the ODE (2.12-2) for $a(t)$. The solutions are

$$\frac{1}{2}\{a^2(0) - a^2(t)\} = \lambda_1 t.$$

Given $a(0) > 0$, we get total collapse at time $t = \frac{a^2(0)}{2\lambda_1}$. The number of particles in R is

$$N = \int_0^a \frac{a}{r}U(\frac{r}{a})(4\pi r^2) \, dr = 4\pi a^3 \int_0^1 \zeta U \, d\zeta,$$

and with the help of (2.12-2), we find that

$$\dot{N} = 12\pi a^2 \dot{a} \int_0^1 \zeta U \, d\zeta = -\left(12\pi \lambda_1 \int_0^1 \zeta U \, d\zeta\right) a(t),$$

which is proportional to $a(t)$.

The diffusion kernel. The distribution $g(x, t)$ of Brownian particles released from the origin $x = 0$ at time $t = 0$ is called a *diffusion kernel*. The distribution $g(x, t)$ is a solution of the diffusion equation

(2.28)
$$g_t = Dg_{xx}, \quad \text{all } x, \text{ all } t > 0,$$

which satisfies the initial condition

(2.29)
$$g(x, t) \to 0 \quad \text{as } t \to 0, \, x \neq 0 \text{ fixed},$$

and the integral condition

$$(2.30) \qquad\qquad \int_{-\infty}^{\infty} g(x,t)\,dx = 1.$$

The diffusion coefficient D can absorbed by nondimensionalization: Let L and T be any fixed units of length and time such that

$$\frac{DT}{L^2} = 1,$$

and use $\frac{1}{L}$ as the unit of g. The dimensionless equations are (2.28)–(2.30) with $D = 1$. Notice that the dimensionless equations are the *same* for *any* choice of L. In this sense, a physicist would say that this process of Brownian particles spreading from the origin "has no characteristic length". We now discuss the underlying mathematical idea.

Scale invariance of the diffusion kernel. The root mean square displacement of particles is

$$\sqrt{\langle x^2 \rangle} = \sqrt{2Dt},$$

as in (2.3). The average value of g over the range of x with $|x| \leq \sqrt{2Dt}$ should have magnitude $\frac{1}{\sqrt{2Dt}}$, to be consistent with the integral condition (2.30). Hence, if we plot $\sqrt{Dt}\,g$ on the vertical axis and $\frac{x}{\sqrt{Dt}}$ on the horizontal, we expect that the plots for different t should be more or less the same. Actually, there is a deep reason for all these plots with different t to be *one and the same graph*.

Let $g(x,t)$ be a solution of the dimensionless equations (2.28)–(2.30) with $D = 1$. For any positive numbers α, β, γ, define the "scaled" distribution

$$g'(x,t) := \gamma g(\frac{x}{\alpha}, \frac{t}{\beta}).$$

It follows from (2.28)–(2.30) that $g'(x,t)$ satisfies

$$\beta g_t' = \alpha^2 g_{xx}', \quad \text{all } x, \text{ all } t > 0,$$
$$g'(x,t) \to 0 \quad \text{as } t \to 0, \ x \neq 0 \text{ fixed},$$
$$\int_{\infty}^{\infty} g'(x,t)\,dx = \alpha\gamma.$$

These equations for g' are exactly the same as (2.28)–(2.30) for g if $\beta = \alpha^2$ and $\gamma = \frac{1}{\alpha}$. Hence,

$$(2.31) \qquad\qquad g'(x,t) = \frac{1}{\alpha} g(\frac{x}{\alpha}, \frac{t}{\alpha^2})$$

solves (2.28)–(2.30) if $g(x,t)$ does. The "absence of a characteristic length", as discussed previously, is clearly a corollary of this *scale invariance* of the equations (2.28)–(2.30). Physically, we expect that the distribution of Brownian particles released from $x = 0$ at $t = 0$ to evolve in a unique way. Mathematically, we assume that the solution of (2.28)–(2.30) is unique, so $g'(x, t)$ in (2.31) must be the same as $g(x, t)$ for any positive value of α, that is,

$$(2.32) \qquad g(x, t) = \frac{1}{\alpha} g(\frac{x}{\alpha}, \frac{t}{\alpha^2})$$

for all x, all $t > 0$, and $\alpha > 0$. At any given $t > 0$, take $\alpha = \sqrt{t}$, which is proportional to $\sqrt{\langle x^2 \rangle}$. Then (2.32) becomes

$$(2.33) \qquad g(x, t) = \frac{1}{\sqrt{t}} g(\frac{x}{\sqrt{t}}, 1) := \frac{1}{\sqrt{t}} S(\zeta := \frac{x}{\sqrt{t}}).$$

If we restore physical units to the variables, the solution (2.33) becomes

$$(2.34) \qquad g(x, t) = \frac{1}{\sqrt{Dt}} S(\zeta := \frac{x}{\sqrt{Dt}}).$$

It is clear that the plot of $\sqrt{Dt} g$ versus $\zeta = \frac{x}{\sqrt{Dt}}$ is the graph of $S(\zeta)$, independent of t. In geometry, two figures of the same shape but possibly different sizes are called similar. Suppose we have two graphs, one of which can be "morphed" into the other by (possibly different) scalings of both axes. We could call these graphs "similar" in a sense slightly more general than the usual geometry sense. This is the reason for calling (2.34) a *similarity solution*.

We can derive a simple ODE boundary value problem for the shape function $S(\zeta)$ in (2.34) and solve it explicitly. First, we compute from (2.34) that

$$g_t = -\frac{1}{2\sqrt{Dt}} \frac{1}{t} \{S(\zeta) + \zeta S'(\zeta)\},$$

$$Dg_{xx} = \frac{1}{\sqrt{Dt}} \frac{1}{t} S''(\zeta).$$

Substituting these expressions for g_t and Dg_{xx} into the diffusion equation (2.28) gives the ODE

$$S'' + \frac{1}{2}(S + \zeta S') = 0$$

or

$$(2.35) \qquad (S' + \frac{\zeta}{2} S)' = 0.$$

In addition, we convert the initial condition (2.29) and integral condition (2.30) into conditions on $S(\zeta)$. It follows from (2.29) and (2.34) that

$$\frac{1}{\sqrt{Dt}} S(\frac{x}{\sqrt{Dt}}) \to 0$$

as $t \to 0$ with $x \neq 0$ fixed, so

(2.36) $\zeta S(\zeta) \to 0$

as $|\zeta| \to \infty$. Substituting (2.34) into the integral condition (2.30) gives

(2.37) $\int_{-\infty}^{\infty} S(\zeta) \, d\zeta = 1.$

It follows from the ODE (2.35) that

$$S' + \frac{\zeta}{2} S = C = \text{constant},$$

or

$$(e^{\frac{\zeta^2}{4}} S)' = C e^{\frac{\zeta^2}{4}},$$

and integration gives

(2.38) $S(\zeta) = S(0) e^{-\frac{\zeta^2}{4}} + C \int_0^{\zeta} e^{\frac{1}{4}(\zeta^2 - \zeta'^2)} \, d\zeta'.$

What is C? The integral term in (2.38) is asymptotic to $\frac{2C}{\zeta}$ as $|\zeta| \to \infty$, that is,

$$C \int_0^{\zeta} e^{\frac{1}{4}(\zeta^2 - \zeta'^2)} \, d\zeta' = \frac{2C}{\zeta} (1 + \varepsilon(\zeta)),$$

where $\varepsilon(\zeta) \to 0$ as $|\zeta| \to \infty$. Hence,

$$\zeta S(\zeta) \to 2C$$

as $|\zeta| \to \infty$, and it follows from (2.36) that $C = 0$. The value of $S(0)$ follows from substituting the $C = 0$ solution (2.38) into (2.37). We get $S(0) = \frac{1}{2\sqrt{\pi}}$. Hence,

$$S(\zeta) = \frac{1}{2\sqrt{\pi}} e^{-\frac{\zeta^2}{4}},$$

and the similarity solution (2.34) takes the explicit form

(2.39) $g(x, t) = \frac{1}{2\sqrt{\pi Dt}} e^{-\frac{x^2}{4Dt}}.$

This is the *diffusion kernel*. It appears in the solution of initial and boundary value problems associated with the diffusion PDE. For instance, in the general integral identity (2.7), which gives $c(x, t+\tau)$ in terms of $c(x, t)$, replace t by 0 and τ by t, and use the explicit form (2.39) of g to get

(2.40) $c(x, t) = \frac{1}{2\sqrt{\pi Dt}} \int_{-\infty}^{\infty} c(x - \zeta, 0) e^{-\frac{\zeta^2}{4Dt}} \, d\zeta.$

This is the solution of the diffusion PDE in $t > 0$, for all x, starting from known values of $c(x, t = 0)$.

Problem 2.13 ($\mathbb{R}^2$ diffusion kernel from symmetry and independence).
The $\mathbb{R}^2$ diffusion kernel $g_2(\boldsymbol{x}, t)$ is examined under the following hypotheses:

a) It is circularly symmetric, that is,

$$g_2(\boldsymbol{x}, t) = G(r := |\boldsymbol{x}|, t).$$

b) Due to statistical independence of $\boldsymbol{e}_1$ and $\boldsymbol{e}_2$ displacements, $g_2(\boldsymbol{x}, t)$ is a *product* of one-dimensional diffusion kernels,

$$g_2(\boldsymbol{x}, t) = g(x_1, t)g(x_2, t)$$

where

$$\int_{-\infty}^{\infty} g \, dx = 1, \quad \int_{-\infty}^{\infty} xg \, dx = 0, \quad \int_{-\infty}^{\infty} x^2 g \, dx = 2Dt.$$

Compute $g_2(\boldsymbol{x}, t)$ from the information in Parts a and b. *Do not* use the explicit solution for the one-dimensional diffusion kernel.

Solution. We have

(2.13-1) $$G(\sqrt{x_1^2 + x_2^2}, t) = g(x_1, t)g(x_2, t).$$

Setting $x_2 = 0$ and $x_1 = x \geq 0$, this becomes

(2.13-2) $$G(x, t) = g(x, t)g(0, t).$$

For $x_1, x_2 \geq 0$ it follows from (2.13-2) that

$$G(x_1, t)G(x_2, t) = g^2(0, t)g(x_1, t)g(x_2, t).$$

By (2.13-1), the right-hand side is $G(0, t)G(\sqrt{x_1^2 + x_2^2}, t)$, so we get a functional equation for G,

(2.13-3) $$G(x_1, t)G(x_2, t) = G(0, t)G(\sqrt{x_1^2 + x_2^2}, t)$$

for $x_1, x_2 \geq 0$. In (2.13-3), set $x_1 = r \cos \vartheta$ and $x_2 = r \sin \vartheta$, with $r \geq 0$ and $0 \leq \vartheta \leq \frac{\pi}{2}$. We get

(2.13-4) $$G(r \cos \vartheta, t)G(r \sin \vartheta, t) = G(0, t)G(r, t).$$

The left-hand side must be independent of ϑ. Differentiate (2.13-4) twice with respect to ϑ and then set $\vartheta = 0$. We get

(2.13-5) $$G(r, t)r^2 \partial_{11} G(0, t) - G(0, t)r\partial_1 G(r, t) = 0.$$

Here, ∂_1 denotes the partial derivative with respect to the first argument of G. We recast (2.13-5) as

$$\partial_1 G(r, t) = -B(t)rG(r, t),$$

where $B(t) := -\partial_{11} G(0, t)/G(0, t)$. The solutions of this ODE for G are

(2.13-6) $$G(r, t) = A(t)e^{-\frac{B(t)}{2}r^2},$$

where $A = A(t)$ is independent of r. The values of $A(t)$ and $B(t)$ follow from moments of $G(r,t)$: From (2.13-1) and the known moments of $g(x,t)$,

$$\int_{\mathbb{R}^2} G(r,t)\, d\boldsymbol{x} = \left(\int_{-\infty}^{\infty} g(x,t)\, dx \right)^2 = 1,$$

$$\int_{\mathbb{R}^2} r^2 G(r,t)\, d\boldsymbol{x} = 2 \int_{-\infty}^{\infty} g(x,t)\, dx \int_{-\infty}^{\infty} x^2 g(x,t)\, dx = 4Dt.$$

Substituting (2.13-6) into these moment identities gives

$$A = \frac{1}{4\pi Dt}, \quad B = \frac{1}{2Dt},$$

and thus the $\mathbb{R}^2$ diffusion kernel is

$$g_2(\boldsymbol{x},t) = G(r,t) = \frac{1}{4\pi Dt} e^{-\frac{r^2}{4Dt}}.$$

Problem 2.14 (Similarity solution for melting interfaces). The free boundary problem for a planar solid-liquid interface is

$$h_t = Dh_{xx} \quad \text{in } x > X(t),$$

$$h = 0, \ Dh_x = -H\dot{X} \quad \text{on } x = X(t).$$

Determine the similarity solution which results from the initial condition

$$X(0) = 0, \ h(x,0) \equiv h_\infty > -H \quad \text{in } x > 0.$$

Give an exact formula for the time to melt a slab of thickness $2a$.

Solution. The effect of heat transfer is felt at distance $\sqrt{Dt}$ from the interface. An order-of-magnitude balance in the heat transfer boundary condition $Dh_x = -H\dot{X}$ on $x = X(t)$ gives $H\dot{X} \simeq \frac{Dh_\infty}{\sqrt{Dt}}$, so $X(t) \simeq \frac{h_\infty}{H}\sqrt{Dt}$. Hence, we seek the similarity solution in the form

$$h(x,t) = h_\infty S(\zeta := \frac{x}{\sqrt{Dt}}) \quad \text{in } \zeta > \zeta_0,$$

$$X(t) = \zeta_0 \sqrt{Dt}.$$

Substituting these representations of $X(t)$ and $h(x,t)$ into the free boundary problem gives an ODE initial value problem for $S(\zeta)$,

(2.14-1) $$S'' + \frac{\zeta}{2} S' = 0 \quad \text{in } \zeta > \zeta_0,$$

(2.14-2) $$S(\zeta_0) = 0, \quad S'(\zeta_0) = -\frac{H}{h_\infty} \frac{\zeta_0}{2}.$$

In addition, we expect $h(x,t) \to h_\infty$ as $x \to +\infty$, which implies a boundary condition

$$S(+\infty) = 1.$$

It follows from this boundary condition that $h(x,0) \equiv h_\infty$ in $x > 0$. Integration of the ODE (2.14-1) and use of the second boundary condition in (2.14-2) gives

(2.14-3)
$$S'(\zeta) = -\frac{H}{h_\infty}\frac{\zeta_0}{2}e^{-\frac{\zeta^2 - \zeta_0^2}{4}}.$$

Integration of (2.14-3) over $\zeta > \zeta_0$ and use of the boundary conditions $S(\zeta_0) = 0$ and $S(+\infty) = 1$ gives

$$1 = -\frac{H}{h_\infty}\frac{\zeta_0}{2}\int_{\zeta_0}^{\infty} e^{-\frac{\zeta^2 - \zeta_0^2}{4}}\,d\zeta$$

or

(2.14-4)
$$f(\zeta_0) := \frac{\zeta_0}{2}\int_{\zeta_0}^{\infty} e^{-\frac{\zeta^2 - \zeta_0^2}{4}} = -\frac{h_\infty}{H}.$$

This is an algebraic equation for ζ_0. The function $f(\zeta_0)$ has the asymptotic behaviors

$$f(\zeta_0) \sim \begin{bmatrix} \frac{\zeta_0}{\sqrt{\pi}}e^{\frac{\zeta_0^2}{4}} & \text{as} & \zeta_0 \to -\infty, \\ \frac{\zeta_0}{2\sqrt{\pi}} & \text{as} & \zeta_0 \to 0, \\ 1 & \text{as} & \zeta_0 \to +\infty. \end{bmatrix}$$

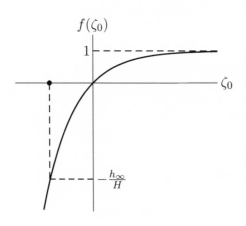

Figure 2.5.

Here the symbol $\sim$ ("is asymptotic to") means that the relative errors of the approximations vanish in the indicated limits. Figure 2.5 shows the graphical solution of (2.14-4). For any $h_\infty > 0$, we get a unique, negative ζ_0, corresponding to a melting interface. For $-H < h_\infty < 0$, we get $\zeta_0 > 0$, corresponding to a freezing interface. Notice the peculiar result, that $\zeta_0 \to +\infty$ as $h_\infty \to -H$ from above. This raises the question: what happens for $h_\infty \leq -H$?

Let us return to the problem of the melting slab. Recall that a slab of thickness $2a$ is immersed in liquid that has $h \equiv h_\infty$ at time zero, with its center plane at $x = 0$. Consider the liquid to the right of the slab: We have $h(x,0) \equiv h_\infty$ in $x - a > 0$. Hence, we expect that h in the liquid to the right evolves according to the similarity solution, with $x - a$ replacing x; that is, $h(x, t) = h_\infty S(\frac{x-a}{\sqrt{Dt}})$ in $x > a - |\zeta_0|\sqrt{Dt}$. Here, we wrote $-|\zeta_0|$ for ζ_0 since $\zeta_0 < 0$. By symmetry, the liquid to the left of the slab has

$$h(x, t) = h_\infty S(\frac{-a - x}{\sqrt{Dt}}) \quad \text{in } x < -a + |\zeta_0|\sqrt{Dt}.$$

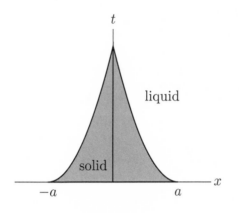

Figure 2.6.

The right and left sides of the slab meet at $x = 0$ at time

(2.14-5) $$t = \frac{1}{\zeta_0^2}\frac{a^2}{D},$$

and this is when the slab is completely melted. In (2.14-5) ζ_0^2 is a function of $\frac{h_\infty}{H}$ as follows from (2.14-4). In the limit of $\frac{h_\infty}{H} \to 0$, $\zeta_0^2 \sim 4\pi\left(\frac{h_\infty}{H}\right)^2$ and the melting time reduces to $\frac{1}{4\pi}\left(\frac{H}{h_\infty}\right)^2\frac{a^2}{D}$, which reproduces the "scaling estimate" (2.27) modulo the prefactor $\frac{1}{4\pi}$.

Problem 2.15 (Source of Brownian particles in two dimensions).
In Problem 2.8, we considered Brownian particles in the exterior of a disk of radius a in $\mathbb{R}^2$. The initial concentration is $c(\boldsymbol{x}, 0) \equiv c_0 = \text{constant}$ in $r > a$, and the circle $r = a$ is a source of Brownian particles, introducing them into $r > a$ at rate S per unit length. We showed that $c(r, t)$ cannot relax to a time-independent limit as $t \to \infty$. Determine an approximate solution for $c(r, t)$ in the limit $t \gg \frac{a^2}{D}$ with $\frac{r}{\sqrt{Dt}}$ fixed.

Solution. Without loss of generality, we assume $c_0 = 0$, so that $c(r, t)$ represents the concentration of particles introduced after time 0. (We can include c_0 as an additive constant afterwards.) The total number of particles at time t is $2\pi aSt$. Hence, we have the integral condition

(2.15-1) $$\int_a^\infty c(r, t)2\pi r\, dr = 2\pi aSt.$$

At times $t \gg \frac{a^2}{D}$, most particles have diffused distance $\sqrt{Dt} \gg a$, so we introduce $\zeta := \frac{r}{\sqrt{Dt}}$ as an independent variable in place of r. The integral condition (2.15-1) translates into

$$\int_{\frac{a}{\sqrt{Dt}}}^{\infty} c(\zeta, t)\zeta \, d\zeta = \frac{aS}{D},$$

and in the limit $t \gg \frac{a^2}{D}$, this reduces to

(2.15-2)
$$\int_0^{\infty} c(\zeta, t)\zeta \, d\zeta = \frac{aS}{D}.$$

The diffusion PDE in ζ, t variables is

(2.15-3)
$$c_t - \frac{\zeta}{2}c_\zeta = c_{\zeta\zeta} + \frac{1}{\zeta}c_\zeta.$$

The absence of explicit time dependence in the $t \gg \frac{a^2}{D}$ integral condition (2.15-2) suggests that the time dependence of c can be ignored for $t \gg \frac{a^2}{D}$, and so (2.15-3) reduces to the ODE

$$c_{\zeta\zeta} + \left(\frac{\zeta}{2} + \frac{1}{\zeta}\right)c_\zeta = 0.$$

Integration gives

(2.15-4)
$$c_\zeta = -\frac{C}{\zeta}e^{-\frac{\zeta^2}{4}},$$

where C is a constant of integration. Since $c \to 0$ as $\zeta \to \infty$, we integrate once more to get

(2.15-5)
$$c = C\int_\zeta^{\infty} \frac{e^{-\frac{s^2}{4}}}{s} \, ds.$$

It remains to choose the constant C so that the integral condition (2.15-2) is satisfied. An integration by parts applied to the left-hand side of (2.15-2) gives

$$\left[c\frac{\zeta^2}{2}\right]_0^{\infty} - \int_0^{\infty} c_\zeta \frac{\zeta^2}{2} \, d\zeta = \frac{aS}{D}.$$

Substituting (2.15-4) for c_ζ, this becomes

$$C\int_0^{\infty} \frac{\zeta}{2}e^{-\frac{\zeta^2}{4}} \, d\zeta = \frac{aS}{D}.$$

The integral on the left-hand side is elementary; its value is 1, so $C = \frac{aS}{D}$. Hence, (2.15-5) becomes

$$c = \frac{aS}{D}\int_\zeta^{\infty} \frac{e^{-\frac{s^2}{4}}}{s} \, ds.$$

Problem 2.16 (Homogeneous harmonic function in a wedge).

a) Let $u(\boldsymbol{x})$ be a smooth function in the wedge W in $\mathbb{R}^2$ between the x-axis and the ray with angle θ from the origin, where $0 < \theta < 2\pi$. We say that $u(\boldsymbol{x})$ is *homogeneous* if there is a smooth function $b = b(a)$ such that

$$(2.16\text{-}1) \qquad\qquad u(a\boldsymbol{x}) = b(a)u(\boldsymbol{x})$$

for all $\boldsymbol{x}$ in W and any positive constant a. Show that there is a constant c so that

$$(2.16\text{-}2) \qquad\qquad u = r^c S(\vartheta)$$

where (r, ϑ) are polar coordinates of $\mathbb{R}^2$ with $0 < \vartheta < \theta$.

b) Compute the homogeneous harmonic functions in the wedge with zero boundary conditions on $\vartheta = 0$ and $\vartheta = \theta$.

Solution.

a) In polar coordinates, the homogeneity property (2.16-1) reads

$$(2.16\text{-}3) \qquad\qquad u(ar, \vartheta) = b(a)u(r, \vartheta).$$

Differentiating (2.16-3) with respect to a and then setting $a = 1$ gives

$$ru_r(r, \vartheta) = b'(1)u(r, \vartheta)$$

or

$$\frac{u_r(r, \vartheta)}{u(r, \vartheta)} = \frac{b'(1)}{r}.$$

The solution of this ODE in r can be written as

$$u(r, \vartheta) = u(1, \vartheta)r^{b'(1)}.$$

Now set $S(\vartheta) := u(1, \vartheta)$ and $c := b'(1)$, so

$$u(r, \vartheta) = r^c S(\vartheta).$$

b) Substitute (2.16-2) into Laplace's equation:

$$0 = \Delta u = u_{rr} + \frac{1}{r}u_r + \frac{1}{r^2}u_{\vartheta\vartheta} = \{c^2 S(\vartheta) + S''(\vartheta)\}r^{c-2}.$$

Hence, we have the ODE for $S(\vartheta)$,

$$S''(\vartheta) + c^2 S(\vartheta) = 0,$$

subject to zero boundary conditions, $S(0) = S(\theta) = 0$. This is a standard ODE eigenvalue problem, and the admissible c's are

$$c = \pm\frac{n\pi}{\theta}$$

with n a positive integer. For given n, the corresponding $S(\vartheta)$ is a multiple of

$$\sin \frac{n\pi}{\theta} \vartheta.$$

In summary, the homogeneous harmonic functions which vanish on $\vartheta = 0, \theta$ are (modulo multiplicative constants)

$$u = r^{\frac{n\pi}{\theta}} \sin \frac{n\pi}{\theta} \vartheta, \ r^{-\frac{n\pi}{\theta}} \sin \frac{n\pi}{\theta} \vartheta,$$

where n ranges over positive integers.

Guide to bibliography. Recommended references for this chapter are Barenblatt [1], Batchelor [2], Chorin & Hald [5], Churchill & Brown [6], Feynman [12, 13], Landau & Lifshitz [16], and Wax [23].

The standard derivation of diffusion PDEs from Fick's law can be found in many of these references [1, 2, 6, 13]. This is the opposite of our discussion, which makes the progression from Brownian motion to the diffusion PDE to Fick's law. In our discussion of Brownian motion, the terms homogeneity and isotropy are used in a sense closely analogous to that in Landau & Lifshitz's [16] discussion of spacetime symmetries in elementary mechanics. Feynman's discussion [12] of why the mean square displacement of Brownian motion is proportional to time is mathematically simple but physically deep. In particular, he derives the Einstein relation between the diffusion and drag coefficients, which we will examine in Chapter 3. Chorin & Hald [5] give a simple presentation of the lattice model of Brownian motion, as well as a Chapman-Kolmogorov derivation of the diffusion PDE like ours. Wax [23] has historical, classic expositions on Brownian motion, but it is easier to read Chorin & Hald [5] or Feynman [12]. The derivation of the diffusion kernel as a similarity solution is presented by Barenblatt [1]. In fact, dimensional analysis, scaling, and scale invariance are the essential subjects of Barenblatt's book, which represents the next level beyond the heuristic treatments given in our book. Problem 2.15 on Brownian particles in $\mathbb{R}^2$ introduced from the perimeter of a disk is an example of "intermediate asymptotics". Problem 2.16 on homogeneous harmonic functions in a wedge is discussed in Barenblatt [1]. The main point is that the scaling of the solutions with r does *not* follow from dimensional analysis.

Chapter 3

Local conservation laws

Definition. The convection PDE (1.8) and the diffusion PDE (2.14) both take the form

(3.1) $$c_t + \boldsymbol{\nabla} \cdot \boldsymbol{f} = 0$$

where the *flux* vector field $\boldsymbol{f}$ is

(3.2) $$\boldsymbol{f} = c\boldsymbol{u}, \quad \boldsymbol{u} = \boldsymbol{u}(\boldsymbol{x}, t) = \text{given velocity field}$$

for convection, or

(3.3) $$\boldsymbol{f} = -D\boldsymbol{\nabla} c$$

for diffusion. (3.1) represents a rather general framework of PDEs in continuum mechanics. Typically, there is a set of *state variables* which can be combined into an equivalent number of densities. Each density satisfies (3.1) with its flux determined by pointwise properties of the original state variables. This set of equations is called a system of *local conservation laws*. The determinations of fluxes from pointwise properties of the state variables are called *constitutive relations*. (3.2) and (3.3) are the constitutive relations for convection and diffusion, respectively. Examples of local conservation laws from real continuum mechanics appear in Chapters 6 and 7. For now, here is another simple example.

Convection-diffusion. Suppose that particles suspended in a fluid are buffeted by the fluid molecules (the same process which by itself leads to diffusion), but in addition, the fluid flows with velocity field $\boldsymbol{u}(\boldsymbol{x}, t)$. Let us look at the particles in a small neighborhood of a material point $\boldsymbol{x}(t)$ of the fluid. This neighborhood is sufficiently small that $\boldsymbol{u}$ is nearly uniform throughout it. In the rest frame of our material point, the particles appear to undergo ordinary diffusion. The rate at which particles cross a small material area $\boldsymbol{n}\, da$ containing $\boldsymbol{x}(t)$ in the direction *opposite* $\boldsymbol{n}$ is $-(-D\boldsymbol{\nabla}c)\cdot\boldsymbol{n}\, da = D\boldsymbol{\nabla}c\cdot\boldsymbol{n}\, da$.

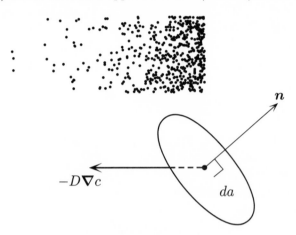

Figure 3.1.

Now consider the influx of particles across the boundary of a macroscopic material region $R(t)$. The diffusion influx process described in Figure 3.1 is happening on each little patch da of $\partial R(t)$, so we get

$$(3.4) \qquad \frac{d}{dt}\int_{R(t)} c\, d\boldsymbol{x} = \int_{\partial R(t)} D\boldsymbol{\nabla}c\cdot\boldsymbol{n}\, da.$$

Applying the transport identity (1.4) to the left-hand side and the divergence theorem to the right-hand side, (3.4) becomes

$$\int_{R(t)} \{c_t + \boldsymbol{\nabla}\cdot(c\boldsymbol{u} - D\boldsymbol{\nabla}c)\}\, d\boldsymbol{x} = 0,$$

and the resulting PDE for $c(\boldsymbol{x},t)$ is the *convection-diffusion* PDE

$$(3.5) \qquad c_t + \boldsymbol{\nabla}\cdot(c\boldsymbol{u} - D\boldsymbol{\nabla}c) = 0.$$

The flux for convection-diffusion is just the sum of the convection and diffusion components.

Problem 3.1 (A "dissolving" interface).   The solid phase of a substance is surrounded by liquid solvent, which has "monomers" (molecular units of the substance) dissolved in it. The volume per monomer in the solid phase is V, and the concentration $c(\boldsymbol{x},t)$ of monomers in the solution is *dilute*, that is, $c(\boldsymbol{x},t) \ll \frac{1}{V}$. The monomer transport in the solution is convection-diffusion.

The diffusion coefficient is D. The velocity field $\boldsymbol{u}(\boldsymbol{x},t)$ of the solvent arises because the solution fills in the space surrounded by the dissolving solid. The solvent liquid is incompressible, and in the dilute limit with small volume fraction of dissolved solid, we have $\boldsymbol{\nabla} \cdot \boldsymbol{u} = 0$. Also, the solvent liquid does not penetrate into the solid, so there is the "no penetration" boundary condition: On the interface, the normal component $\boldsymbol{u} \cdot \boldsymbol{n}$ of liquid velocity equals v, the normal velocity of the interface.

a) Show that $c(\boldsymbol{x},t)$ satisfies the conservation boundary condition $Dc_n = \frac{v}{V}$ on the interface.

b) To appreciate the role of the "displacement" velocity field $\boldsymbol{u}$, let us consider another process, from which it is absent: We have monomers on a planar surface. The analogy of the "solid" in Part a is a densely packed monolayer with area A per monomer. We call regions occupied by this monolayer phase "islands". The islands are surrounded by a dilute two-dimensional "gas" of monomers undergoing diffusion, presumably induced by thermal agitation of the planar substrate. Let $c(\boldsymbol{x},t)$ be the concentration of monomers in the "gas". Derive the conservation boundary condition that applies to c on the island-gas interface.

Solution.

a) Let $\bar{B}(t)$ be a finite region occupied by the solid phase, while $B := \mathbb{R}^3 - \bar{B}$ is occupied by solution. Conservation of monomers is expressed by

$$(3.1\text{-}1) \qquad \frac{d}{dt} \int_{\bar{B}} \frac{d\boldsymbol{x}}{V} + \frac{d}{dt} \int_B c(\boldsymbol{x},t)\,d\boldsymbol{x} = 0.$$

Let $\boldsymbol{n}$ and v denote the unit normal and normal velocity of ∂B. Remember that the orientation of $\boldsymbol{n}$ is *outward* from B, and that v is positive when B "advances" into space formerly occupied by $\bar{B}$. We have

$$(3.1\text{-}2) \qquad \frac{d}{dt} \int_{\bar{B}} \frac{d\boldsymbol{x}}{V} = - \int_{\partial B} \frac{v}{V}\,da$$

and

$$(3.1\text{-}3) \qquad \frac{d}{dt} \int_B c(\boldsymbol{x},t)\,d\boldsymbol{x} = \int_B c_t\,d\boldsymbol{x} + \int_{\partial B} vc\,da.$$

In (3.1-3), $v = \boldsymbol{u} \cdot \boldsymbol{n}$ on ∂B, and we use the divergence theorem to convert the surface integral into a volume integral. Thus, (3.1-3) becomes

$$\frac{d}{dt} \int_B c\,d\boldsymbol{x} = \int_B \{c_t + \boldsymbol{\nabla} \cdot (c\boldsymbol{u})\}\,d\boldsymbol{x}.$$

By the convection-diffusion PDE (3.5), $c_t + \nabla \cdot (c\boldsymbol{u}) = D\Delta c$, and use of the divergence theorem gives

(3.1-4) $$\frac{d}{dt} \int_B c\,d\boldsymbol{x} = \int_{\partial B} Dc_n\,da.$$

From (3.1-1), (3.1-2) and (3.1-4), it follows that

$$\int_{\partial B} \{Dc_n - \frac{v}{V}\}\,da = 0.$$

The integrand represents the "rate at which particles are generated per unit area on ∂R", which is of course zero, so

$$Dc_n = \frac{v}{V}$$

on ∂B.

b)   Let $\bar{B}$ denote an "island", and $B := \mathbb{R}^2 - \bar{B}$ the surrounding "gas". Equations (3.1-1)–(3.1-3) from Part a apply here, with A replacing V. But now, substitute $c_t = D\Delta c$ in (3.1-3) and use the divergence theorem. Then (3.1-3) becomes

(3.1-5) $$\frac{d}{dt} \int_B c\,d\boldsymbol{x} = \int_{\partial B} \{Dc_n + vc\}\,da.$$

It now follows from (3.1-1), (3.1-2) and (3.1-5) that

$$\int_{\partial B} \{Dc_n - (\frac{1}{A} - c)v\}\,da = 0,$$

and the boundary condition is

$$Dc_n = \frac{v}{A} - cv$$

on ∂B. To understand the term $-cv$, which did not appear in the boundary condition of Part a, consider the case of a "receding" island ($v > 0$). In the rest frame of the interface, we see island particles convecting to the interface at a rate of $\frac{v}{A}$ per unit length, and there is a convective flux cv of "gas" particles away from the interface. The diffusive flux Dc_n accounts for the difference $\frac{v}{A} - cv$.

Smoluchowski boundary value problem. In liquid solution, there is another class of convection-diffusion processes: The liquid induces diffusion of particles, with diffusion coefficient D as before. In addition, the particles are subject to an "external force" $-\nabla\varphi(\boldsymbol{x})$ which derives from a potential energy field $\varphi(\boldsymbol{x})$. A simple example would be the buoyancy force on particles in a mud puddle. Assuming the liquid is at rest, the external force induces an average "drift" of particles with *drift velocity* $\boldsymbol{u}$ given by

$$(3.6) \qquad\qquad \zeta\boldsymbol{u} = -\nabla\varphi.$$

The positive constant ζ is called the *drag* coefficient of the particles: The drift velocity in (3.6) results from a balance between the external force and a *drag force* equal to $-\zeta\boldsymbol{u}$. One deep point of this Smoluchowski "drift-diffusion" model of particle transport is the relationship between ζ and D,

$$(3.7) \qquad\qquad \zeta = \frac{k_B T}{D}.$$

Here, T is the temperature and k_B is the Boltzmann constant, so that $k_B T$ is the characteristic kinetic energy of random thermal motions at the molecular level. Once $k_B T$ is recognized as an energy, the relation (3.7) is suggested by dimensional analysis. But how does it arise fundamentally?

The transport of particles in a region R due to the combined effects of drift and diffusion is modeled by a convection-diffusion PDE for the particle concentration $c(\boldsymbol{x}, t)$. The PDE is (3.1) with the flux

$$(3.8) \qquad\qquad \boldsymbol{f} = -\frac{1}{\zeta}c\nabla\varphi - D\nabla c.$$

If the region R is a "container" with impenetrable boundary ∂R, we have the boundary condition

$$(3.9) \qquad\qquad 0 = -\boldsymbol{f}\cdot\boldsymbol{n} = \frac{1}{\zeta}c\varphi_n + Dc_n$$

on ∂R. In Problem 3.2 it is shown that *all* time-independent solutions of the boundary value problem (3.1)–(3.9) for c are proportional to

$$(3.10) \qquad\qquad e^{-\beta\varphi(\boldsymbol{x})},$$

where $\beta := \frac{1}{D\zeta}$. The exponential (3.10) is reminiscent of a basic result from equilibrium thermodynamics: The thermodynamic equilibrium of our particles in solution is characterized by $c = c(\boldsymbol{x})$ independent of time and proportional to the *Boltzmann factor*

$$(3.11) \qquad\qquad e^{-\frac{\varphi(\boldsymbol{x})}{k_B T}}.$$

Comparing (3.10) and (3.11) leads to the identification $\beta = \frac{1}{D\zeta} = \frac{1}{k_B T}$, and the relation (3.7) between ζ and D follows.

If we simply take the Boltzmann factor (3.11) as given, the "derivation" of (3.7) is facile. What makes (3.7) "deep" is the thermodynamic origin of the Boltzmann factor, which is independent of the convection-diffusion model. The thermodynamic idea is that the equilibrium $c(\boldsymbol{x})$ minimizes the *free energy* functional

$$(3.12) \qquad f[c] := \int_R \left\{ \varphi c + k_B T \big(c \log(vc) - c \big) \right\} d\boldsymbol{x}$$

over the function space of positive $c(\boldsymbol{x})$ corresponding to a fixed number N of particles:

$$(3.13) \qquad \int_R c \, d\boldsymbol{x} = N.$$

In (3.12), v is a positive constant with the units of volume, so vc is dimensionless. The integral $\int_R \varphi c \, d\boldsymbol{x}$ is clearly the total potential energy of all the particles. The integral

$$(3.14) \qquad S := -\int_R \left\{ c \log(vc) - c \right\} d\boldsymbol{x},$$

called *entropy*, measures the "number of microscopic configurations consistent with a given concentration field $c(\boldsymbol{x})$". The precise sense, based on combinatorics, is spelled out in Problem 3.3. The decrease of free energy can be accomplished by decreasing the potential energy or increasing the entropy. Hence, equilibrium represents a compromise between the tendency of the particle system to "settle into a stable low-energy configuration" and the tendency to "spread indiscriminately throughout all possible microscopic configurations" due to the thermal buffeting.

We show how the Boltzmann factor arises from minimization of free energy. Let $c = c(\boldsymbol{x}, \varepsilon)$ be a family of concentration fields parametrized by ε, such that the particle number constraint (3.13) holds for each ε and the minimum free energy is achieved for $\varepsilon = 0$. Substituting $c = c(\boldsymbol{x}, \varepsilon)$ into the free energy (3.12) makes it a function of ε, and a necessary condition for it to have a minimum at $\varepsilon = 0$ is

$$(3.15) \qquad 0 = \frac{df}{d\varepsilon}(0) = \int_R \left\{ \varphi + k_B T \log\big(vc(\boldsymbol{x}, 0) \big) \right\} c_\varepsilon(\boldsymbol{x}, 0) \, d\boldsymbol{x}$$

for all $c_\varepsilon(\boldsymbol{x}, 0)$ such that

$$(3.16) \qquad \int_R c_\varepsilon(\boldsymbol{x}, 0) \, d\boldsymbol{x} = 0.$$

The latter integral constraint on $c_\varepsilon(\boldsymbol{x}, 0)$ follows from ε-differentiation of

$$\int_R c(\boldsymbol{x}, \varepsilon) \, d\boldsymbol{x} = N.$$

Now take any two points x_1 and x_2 in R and $a > 0$ sufficiently small that the disks

$$D_1 : |x - x_1| < a \quad \text{and} \quad D_2 : |x - x_2| < a$$

are in R and disjoint. Take $c_\varepsilon(x, 0)$ to be $\frac{1}{\pi a^2}$ in D_1, $-\frac{1}{\pi a^2}$ in D_2, and zero in the rest of R. Then the constraint (3.16) is satisfied, and taking the $a \to 0$ limit of (3.15) gives

$$\left[\varphi(x) + k_B T \log\big(vc(x, 0)\big)\right]_{x_2}^{x_1} = 0.$$

Hence,

$$\varphi(x) + k_B T \log\big(vc(x, 0)\big)$$

is a uniform constant in R, and it follows that $c(x, 0)$ is proportional to the Boltzmann factor (3.11).

There is more: Let $c(x, t)$ be a time-dependent solution of the Smoluchowski boundary value problem (3.1)–(3.9). In Problem 3.4 it is shown that the free energy (3.12) is *decreasing* in time whenever the flux f in (3.8) is not identically zero; so solutions of the Smoluchowski boundary value problem relax to thermodynamic equilibria.

Problem 3.2 (Time-independent equilibria). Show that any time-independent solution of the Smoluchowski boundary value problem is proportional to $e^{-\beta\varphi(x)}$.

Solution. Write the flux (3.8) as

$$f = -De^{-\beta\varphi}\nabla(e^{\beta\varphi}c).$$

Introducing

(3.2-1) $$g := e^{\beta\varphi}c$$

as the dependent variable in place of c, we have

$$f = -De^{-\beta\varphi}\nabla g.$$

Hence, the time-independent boundary value problem (3.1)–(3.9) can be written as

(3.2-2)
$$\nabla \cdot \left(e^{-\beta\varphi}\nabla g\right) = 0 \quad \text{in } R,$$
$$g_n = 0 \quad \text{on } \partial R.$$

Certainly, $g \equiv$ constant solves (3.2-2), and then (3.2-1) implies that c is proportional to $e^{-\beta\varphi}$. Here, we need to show that g *must* be a uniform constant: Look at the "divergence identity"

$$\boldsymbol{\nabla} \cdot \left(g e^{-\beta\varphi} \boldsymbol{\nabla} g\right) = g \boldsymbol{\nabla} \cdot \left(e^{-\beta\varphi} \boldsymbol{\nabla} g\right) + e^{-\beta\varphi} |\boldsymbol{\nabla} g|^2.$$

The first term on the right-hand side vanishes by the PDE in (3.2-2). Integrating what remains over R and using the divergence theorem on the left-hand side gives

$$\int_R e^{-\beta\varphi} g g_n \, d\varphi = \int_R e^{-\beta\varphi} |\boldsymbol{\nabla} g|^2 \, d\boldsymbol{x}.$$

The left-hand side vanishes by the boundary condition $g_n = 0$ in (3.2-2). Hence,

$$\int_R e^{-\beta\varphi} |\boldsymbol{\nabla} g|^2 \, d\boldsymbol{x} = 0,$$

and $\boldsymbol{\nabla} g := 0$ in R follows; so g is a uniform constant in R.

Problem 3.3 (Entropy functional).   We address the subtle combinatorics argument that underlies the entropy functional (3.14). We approximate R as the union of $M \geq N$ little cubes of volume v. We think of v as the "particle volume". Each little cube can be empty or occupied by one particle. A "microscopic configuration" is the assignment of the N cubes that are occupied.

a) Give a simple counting argument that the total number of microscopic configurations is $\frac{M!}{(M-N)!N!}$.

b) The answer in Part a is the number of configurations, irrespective of compliance with any preordained concentration field $c(\boldsymbol{x})$. This time, we divide our array of M little cubes into "blocks", each with ΔM cubes, where $1 \ll \Delta M \ll M$. Let $\boldsymbol{x}_j$ be a point inside the jth block. We assign $\Delta N_j := c(\boldsymbol{x}_j) v \Delta M$ particles to the jth block, to comply with the preordained concentration $c(\boldsymbol{x})$. Write down a formula for the total number n of microscopic configurations consistent with the given assignments of ΔN_j particles to the jth block.

c) In statistical mechanics, the *entropy* is defined as the *logarithm* of the number of particle configurations. The idea is that the *composite* of two systems with n_1 and n_2 configurations has $n_1 n_2$ configurations; hence, the entropy of the composite system is $\log(n_1 n_2) = \log n_1 + \log n_2$, the *sum* of the component entropies. Now, given n in Part b, approximate $S := \log n$ as a Riemann sum of an integral involving $c(\boldsymbol{x})$. Here, "approximation" means use of Stirling's approximation $\log(n!) \sim n \log n - n$ for $n \gg 1$. Show that the integral reduces to the entropy in (3.14) in the limit of small volume fraction, $0 \leq vc(\boldsymbol{x}) \ll 1$.

Solution.

a) Place the particles one by one: a choice of M particle cubes for the "first" particle, $M - 1$ for the "second", etc., down to $M - N + 1$ for the "Nth". The total number of these "ordered" assignments of "first", "second", etc. particles is

$$(3.3\text{-}1) \qquad M(M-1)\cdots(M-N+1) = \frac{M!}{(M-N)!}.$$

This number overcounts the number of configurations, because the chosen cubes may be occupied by *any* particles, "first", "second", etc. being irrelevant. For a given configuration of N occupied cubes, there are $N!$ permutations of particles between them. Hence, the total number of distinct configurations is the number (3.3-1) divided by $N!$, or $\frac{M!}{(M-N)!N!}$.

b) The assignment of ΔN_j occupied sites among the ΔM cubes of block j is just a rerun of Part a, and the number of configurations inside block j is $\frac{(\Delta M)!}{(\Delta M - \Delta N_j)!(\Delta N_j)!}$. The total number of configurations for the composite system made of all the blocks is the product

$$n = \prod_j \frac{(\Delta M)!}{(\Delta M - \Delta N_j)!(\Delta N_j)!}.$$

c) Taking the log of n in Part b and using Stirling's approximation, we get

$$S := \log n \sim -\sum_j \left\{ \left(1 - \frac{\Delta N_j}{\Delta M}\right)\log\left(1 - \frac{\Delta N_j}{\Delta M}\right) + \frac{\Delta N_j}{\Delta M}\log\frac{\Delta N_j}{\Delta M} \right\}\Delta M.$$

Now set $\Delta N_j = c(\boldsymbol{x}_j)v\Delta M$ to get

$$S \sim -\frac{1}{v}\sum_j \left\{ (1 - vc(\boldsymbol{x}_j))\log(1 - vc(\boldsymbol{x}_j)) + vc(\boldsymbol{x}_j)\log vc(\boldsymbol{x}_j) \right\}v\Delta M.$$

The right-hand side is a Riemann sum for the integral

$$(3.3\text{-}2) \qquad S \sim -\frac{1}{v}\int_R \left\{ (1 - vc)\log(1 - vc) + vc\log(vc) \right\}d\boldsymbol{x}.$$

In (3.3-2), $vc(\boldsymbol{x})$ is the local volume fraction occupied by particles. In the limit $0 \leq vc \ll 1$, (3.3-2) reduces to

$$S \sim -\int_R \left\{ c\log(vc) - c \right\}d\boldsymbol{x}$$

as in (3.14).

Problem 3.4 (Free energy decreases).   Let $c(\boldsymbol{x}, t)$ be a time-dependent solution of the Smoluchowski boundary value problem (3.1)–(3.9). Show that the free energy (3.12) is decreasing in time whenever the flux $\boldsymbol{f}$ in (3.8) is not identically zero in R.

Solution. From (3.1) and (3.12) we compute

$$\dot{f} = \int_R \{\varphi + k_B T \log(vc)\} c_t \, d\boldsymbol{x}$$

(3.4-1)
$$= -\int_R \{\varphi + k_B T \log(vc)\} \boldsymbol{\nabla} \cdot \boldsymbol{f} \, d\boldsymbol{x}.$$

Next, integrate the "divergence identity"

$$\boldsymbol{\nabla} \cdot \Big((\varphi + k_B T \log(vc)) \boldsymbol{f} \Big) = \Big(\boldsymbol{\nabla}\varphi + k_B T \frac{\boldsymbol{\nabla} c}{c} \Big) \cdot \boldsymbol{f} + (\varphi + k_B T \log(vc)) \boldsymbol{\nabla} \cdot \boldsymbol{f}$$

over R. On the left-hand side use the divergence theorem and the boundary condition $\boldsymbol{f} \cdot \boldsymbol{n} = 0$ on ∂R. We find that $\dot{f}$ in (3.4-1) is given by

(3.4-2)
$$\dot{f} = \int_R \Big(\boldsymbol{\nabla}\varphi + k_B T \frac{\boldsymbol{\nabla} c}{c} \Big) \cdot \boldsymbol{f} \, d\boldsymbol{x}.$$

In the right-hand side observe that

$$\boldsymbol{\nabla}\varphi + k_B T \frac{\boldsymbol{\nabla} c}{c} = \frac{\zeta}{c} \Big(\frac{1}{\zeta} c \boldsymbol{\nabla}\varphi + D\boldsymbol{\nabla} c \Big) = -\frac{\zeta}{c} \boldsymbol{f}.$$

Hence, (3.4-2) becomes

$$\dot{f} = -\zeta \int_R \frac{|\boldsymbol{f}|^2}{c} \, d\boldsymbol{x}.$$

The right-hand side is negative whenever $\boldsymbol{f}$ is not identically zero.

Problem 3.5 (Equilibrium or nonequilibrium in Coulomb potential).

The concentration $c(\boldsymbol{x}, t)$ of particles in the exterior of the disk $r < a$ in $\mathbb{R}^2$ satisfies the Smoluchowski PDE with diffusivity D and $\varphi(\boldsymbol{x}) = E \log \frac{r}{a}$, where E is a positive constant. For $E > 2k_B T$, compute the equilibrium concentration field corresponding to N particles. This corresponds to a "cloud" of particles more or less trapped in an annulus about the disk. If $E < 2k_B T$, it turns out that the force $-\boldsymbol{\nabla}\varphi = \frac{E}{r} \boldsymbol{e}_r$ towards the origin is not strong enough to overcome diffusion and particles tend to disperse to $r = \infty$. There is a similarity solution of the Smoluchowski PDE that represents such a dispersal process in the limit $t \to \infty$ with $\frac{r}{\sqrt{Dt}}$ fixed. Determine it.

Solution. The equilibrium solution takes the form

$$c(\boldsymbol{x}) = c_0 e^{-\frac{\varphi}{k_B T}} = c_0 \left(\frac{r}{a}\right)^{-\frac{E}{k_B T}}.$$

The constant c_0 is determined via the integral constraint

$$N = \int_{r>a} c\,d\boldsymbol{x} = c_0 \int_a^\infty \left(\frac{r}{a}\right)^{-\frac{E}{k_B T}} 2\pi r\,dr = \frac{2\pi a^2 c_0}{\frac{E}{k_B T} - 2},$$

and we get

$$c(\boldsymbol{x}) = \left(\frac{E}{k_B T} - 2\right) \frac{N}{2\pi a^2} \left(\frac{r}{a}\right)^{-\frac{E}{k_B T}}$$

in $r > a$. For $E < 2k_B T$, we seek similarity solutions of the Smoluchowski PDE in the form

$$(3.5\text{-}1) \qquad c = \frac{1}{Dt} S(\zeta := \frac{r}{\sqrt{Dt}}).$$

The prefactor $\frac{1}{Dt}$ upholds normalization. We have

$$\int_{r>a} c\,d\boldsymbol{x} = \int_a^\infty \frac{1}{Dt} S(\zeta) 2\pi r\,dr = 2\pi \int_{\frac{a}{\sqrt{Dt}}}^\infty S(\zeta)\zeta\,d\zeta.$$

In the limit $t \to \infty$, the right-hand side converges to

$$(3.5\text{-}2) \qquad N := 2\pi \int_0^\infty S(\zeta)\zeta\,d\zeta,$$

a constant independent of t. In the circularly symmetric case, the Smoluchowski PDE reads

$$(3.5\text{-}3) \qquad c_t - \frac{DE}{k_B T} \frac{c_r}{r} = D\left(c_{rr} + \frac{1}{r} c_r\right).$$

Substituting (3.5-1) into (3.5-3) gives an ODE for $S(\zeta)$,

$$-\left(S + \frac{\zeta}{2} S'\right) - \frac{E}{k_B T} \frac{S'}{S} = S'' + \frac{1}{\zeta} S',$$

which simplifies to

$$\left\{\zeta S' + \left(\frac{E}{k_B T} + \frac{\zeta^2}{2}\right) S\right\}' = 0.$$

It follows that

$$(3.5\text{-}4) \qquad \zeta S' + \left(\frac{E}{k_B T} + \frac{\zeta^2}{2}\right) S = \text{constant}.$$

We expect S to converge to zero strongly as $\zeta \to \infty$, so the constant should be zero. Integration of (3.5-4) with zero right-hand side gives

$$(3.5\text{-}5) \qquad S(\zeta) = A\zeta^{-\frac{E}{k_B T}} e^{-\frac{\zeta^2}{4}}.$$

The constant A is determined by substituting (3.5-5) into the normalization condition (3.5-2). We have

$$2\pi A \int_0^\infty \zeta^{-(\frac{E}{k_B T} - 1)} e^{-\frac{\zeta^2}{4}} \, d\zeta = N.$$

This time, integrability at $\zeta = 0$ requires $E < 2k_B T$, which is indeed the case we are dealing with here.

Locality. The integral form of (3.1) with respect to a *fixed* region R is

$$(3.17) \qquad \frac{d}{dt} \int_R c \, d\boldsymbol{x} = - \int_{\partial R} \boldsymbol{f} \cdot \boldsymbol{n} \, da.$$

A natural interpretation of (3.17) is that stuff enters R only by crossing the boundary ∂R at a rate per unit area equal to $-\boldsymbol{f} \cdot \boldsymbol{n}$. We say that the transport of stuff is *local* if there is a flux $\boldsymbol{f}$ determined by pointwise properties of its density, such that stuff enters a region R only by crossing the boundary ∂R at a rate $-\boldsymbol{f} \cdot \boldsymbol{n}$ per unit area. Locality is physically reasonable: There is no known instantaneous transport of matter over macroscopic distances, and nor is there instantaneous interaction between distant bodies. Nevertheless, effectively nonlocal transport can arise as an approximation. Here is a simple example.

In an electrically conducting medium, the flux of charge, denoted by $\boldsymbol{j} = \boldsymbol{j}(\boldsymbol{x}, t)$, is called the *current density*. (3.1) applies with c equal to the charge density and with $\boldsymbol{f}$ replaced by the current density $\boldsymbol{j}$. In typical applications, such as electrophysiology, the characteristic time for the relaxation of any imbalance between densities of positive and negative charges is so rapid that a net charge density is not observed. In this case the left-hand side of (3.17) is zero, and

$$\int_{\partial R} \boldsymbol{j} \cdot \boldsymbol{n} \, da = 0$$

for an arbitrary region R. Hence, in regions with no "sources" (i.e., no electrodes connected to batteries), we have

$$(3.18) \qquad \qquad \boldsymbol{\nabla} \cdot \boldsymbol{j} = 0.$$

If the conducting medium is uniform and isotropic, $\boldsymbol{j}(\boldsymbol{x}, t)$ is determined from the *electric potential* $\varphi(\boldsymbol{x}, t)$ by Ohm's law:

$$(3.19) \qquad \qquad \boldsymbol{j} = -\sigma \boldsymbol{\nabla} \varphi.$$

Here, σ is a uniform constant, called the *conductivity* of the medium. It follows from (3.18) and (3.19) that φ is harmonic, i.e.,

(3.20) $$\Delta\varphi = 0.$$

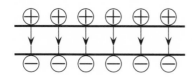

Figure 3.2.

In electrophysiology, the media inside and outside of cells are often modeled as uniform conductors. The cell walls are (relatively) nonconducting, so they are approximated as surfaces with a characteristic *capacitance* C per unit area. Figure 3.2 depicts a small portion of cell wall in its role as capacitor. If the charge per unit area on the upper surface is q, overall electrical neutrality obliges a charge per unit area of $-q$ on the lower. There is a strong electric field inside the cell wall, and the drop in electric potential from the top to the bottom is

(3.21) $$[\varphi] = \frac{q}{C},$$

where C is the capacitance per unit area. We are ready to formulate our example of nonlocal transport: Uniform conducting medium fills $\mathbb{R}^3$, except for an insulating layer about the plane $y = 0$, with capacitance C. We ignore the thickness of this layer, so Laplace's equation (3.20) applies in $\mathbb{R}^3$ minus the $y = 0$ plane, and the electric potential satisfies the jump condition

(3.22) $$[\varphi] := \varphi(x, y = 0^+, z) - \varphi(x, y = 0^-, z) = \frac{q(x, z, t)}{C},$$

where $q(x, z, t)$ is the surface charge density on $y = 0^+$. In addition, we impose the boundary condition

(3.23) $$\varphi \to 0 \quad \text{as} \quad r \to \infty,$$

which says that there are no charges far from the origin.

We quantify the evolution of the surface charge density in the case where φ and q are independent of z. Figure 3.3 is a cartoon of the charge transport in this case. The oriented curves are integral curves of electric field $-\boldsymbol{\nabla}\varphi$, which indicate the current density $\boldsymbol{j}$. Due to overall electrical neutrality, the current is supported by positive charges moving in the direction of the electric field and negative charges moving in the opposite direction. We see that positive charges on $y = 0^+$ in $x_1 < x < x_2$ are lifting into $y > 0$. They "flow" to the right in the shaded region and "land" on $y = 0^+$ in $x'_2 < x < x'_1$. Negative charges are lifting from $x'_2 < x < x'_1$ and landing in $x_1 < x < x_2$. A "mirror image" of this process is happening in $y < 0$. The result is a redistribution of the surface charge $q(x, t)$, but it is clear that charges are not simply moving along the line $y = 0$ according to some

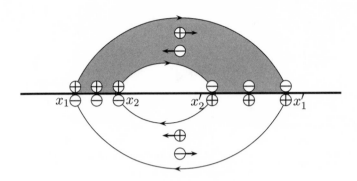

Figure 3.3.

one-dimensional convection-diffusion. So what *is* the evolution equation for
$q(x,t)$?

The total charge on $y = 0^+$ between $x = x_1$ and $x = x_2$ has time rate of
change

$$\frac{d}{dt}\int_{x_1}^{x_2} q(x,t)\,dx = -\int_{x_1}^{x_2} \boldsymbol{j}(x,0^+,t)\cdot \boldsymbol{e}_y\,dx$$

$$= \sigma \int_{x_1}^{x_2} \varphi_y(x,0^+,t)\,dx.$$

This holds for any x_1 and x_2, so we get the local identity

(3.24) $q_t(x,t) = \sigma\varphi_y(x,0^+,t).$

$\varphi_y(x,0^+,t)$ can be expressed in terms of $q(x,t)$. First, the solution of the
boundary value problem (3.20), (3.22) and (3.23) in the z-independent case
is

$$\varphi(x,y,t) = \frac{1}{2\pi C}\int_{-\infty}^{\infty} \frac{y\,q(x',t)}{(x-x')^2 + y^2}\,dx'$$

$$= \frac{1}{4\pi C}\partial_y\int_{-\infty}^{\infty} q(x',t)\ln((x-x')^2 + y^2)\,dx',$$

for $y \neq 0$; hence,

(3.25) $\varphi_y(x,y,t) = \frac{1}{4\pi C}\partial_{yy}\int_{-\infty}^{\infty} q(x',t)\ln((x-x')^2 + y^2)\,dx'.$

In (3.25), $\ln((x-x')^2 + y^2)$ is radially symmetric about $(x',0)$ and solves
Laplace's equation in $(x,y) \neq (x',0)$. The superposition

$$\int_{-\infty}^{\infty} q(x',t)\ln((x-x')^2 + y^2)\,dx$$

solves Laplace's equation in $y \neq 0$. Hence, we may trade ∂_{yy} in (3.25) for
$-\partial_{xx}$ and get

$$\varphi_y(x,y,t) = -\frac{1}{4\pi C}\partial_{xx}\int_{-\infty}^{\infty} q(x',t)\ln((x-x')^2 + y^2)\,dx'.$$

Take one of the ∂_x's inside:

$$\varphi_y(x,y,t) = -\frac{1}{2\pi C}\partial_x \int_{-\infty}^{\infty} q(x',t)\frac{x-x'}{(x-x')^2+y^2}\,dx'.$$

The limit $y \to 0^+$ gives

(3.26) $$\varphi_y(x,0^+,t) = \frac{1}{2\pi C}\partial_x \int_{-\infty}^{\infty} \frac{q(x',t)}{x'-x}\,dx'.$$

The improper integral on the right-hand side is interpreted as a principal value,

$$\int_{-\infty}^{\infty} \frac{q(x',t)}{x'-x}\,dx' := \lim_{\varepsilon \to 0}\int_{|x'-x|>\varepsilon} \frac{q(x',t)}{x'-x}\,dx'.$$

Finally, substitution of (3.26) into (3.24) gives an integro-differential equation for $q(x,t)$,

(3.27) $$q_t(x,t) = \frac{\sigma}{2\pi C}\partial_x\left(\int_{-\infty}^{\infty} \frac{q(x',t)}{x'-x}\,dx'\right).$$

This peculiar equation has a "flux"

$$f = -\frac{\sigma}{2\pi C}\int_{-\infty}^{\infty} \frac{q(x',t)}{x'-x}\,dx'.$$

But as we have seen, it does not arise from a truly "local" transport of charges confined to $y = 0$.

Sources. Suppose some stuff with density $c(\boldsymbol{x},t)$ is transported by a flux $\boldsymbol{f}(\boldsymbol{x},t)$ and, in addition, is "created" at a rate of $s(\boldsymbol{x},t)$ per unit volume. The rate of change of total stuff inside a fixed region R is

(3.28) $$\frac{d}{dt}\int_R c\,d\boldsymbol{x} = -\int_{\partial R} \boldsymbol{f}\cdot\boldsymbol{n}\,da + \int_R s(\boldsymbol{x},t)\,d\boldsymbol{x}.$$

The surface integral represents transport into R across ∂R by the flux $\boldsymbol{f}$. The volume integral is the rate of creation of stuff in R due to the source density s. Assuming that (3.28) holds for all R's, we get the pointwise identity

(3.29) $$c_t + \nabla \cdot \boldsymbol{f} = s.$$

In the realms of continuum mechanics and thermodynamics, we encounter natural sources: Thermal energy in a fluid is transported by convection and diffusion, and can be generated by an exothermic reaction of chemicals in solution. The gravity force per unit volume is a source of momentum density. In the simple context of particles in liquid solution, we cannot have their spontaneous creation from nothing, but one kind of particle can convert into others, as in this next example.

We have a reaction

$$A + B \leftrightarrows C.$$

Let $a(\boldsymbol{x}, t), b(\boldsymbol{x}, t), c(\boldsymbol{x}, t)$ be the concentrations of A, B, C in solution. Suppose that a, b, c are uniform in space, so that they are functions of time only. The concentrations $a(t), b(t), c(t)$ satisfy simple mass action ODEs:

(3.30) $\dot{c} = kab - \ell c,$

(3.31) $\dot{a} = \dot{b} = -kab + \ell c.$

The idea of (3.30) is that each C molecule is created by collision of an A with a B. The collision rate is presumably proportional to the product ab, and k is a creation rate constant. C's can also spontaneously decay into separate A's and B's, with rate constant ℓ. (3.31) indicates that each C is produced at the cost of an A and a B, and the breakup of a C gives an A and a B. Notice that (3.30) and (3.31) imply a conservation of total particles, since $a + b + 2c$ is a constant independent of t. Now suppose that a, b, c depend on spatial position $\boldsymbol{x}$ as well, and that A, B, C are transported by diffusion with diffusion coefficients D_A, D_B, D_C. In place of ODEs (3.30) and (3.31) we get PDEs

$$c_t = D_C \Delta c + kab - \ell c,$$
(3.32) $$a_t = D_A \Delta a - kab + \ell c,$$
$$b_t = D_B \Delta b - kab + \ell c.$$

The sources in these PDEs arise from transmutation of an A and a B into a C and vice versa.

The steady diffusion flame. Assume that the back reaction $C \to A + B$ is negligible so that in (3.32), ℓ is effectively zero. Then the second two PDEs of (3.32) form a closed system for a and b. Assume for simplicity that D_A and D_B have a common value D. Now suppose we have pure A coming from $x = +\infty$, pure B coming from $x = -\infty$, and that there is a "flame" layer about $x = 0$, where the product C is produced. The simplest model of such a flame is based on time-independent one-dimensional solutions of the PDEs for a and b. Under the above simplifying assumptions, $a = a(x)$ and $b = b(x)$ satisfy the ODEs

(3.33) $Da'' = kab, \qquad Db'' = kab.$

As $|x| \to \infty$ we expect the source term kab to vanish. More than that, we expect that the total rate of producing C per unit area of flame,

(3.34) $$\gamma := \int_{-\infty}^{\infty} kab\, dx,$$

is *finite*. Since there is pure B as $x \to -\infty$, we expect $a, a' \to 0$ as $x \to -\infty$. We deduce from the first ODE in (3.33) that $Da'(\infty) = \gamma$. In summary, we have the following boundary conditions on a:

$$(3.35) \qquad\qquad a(-\infty) = 0, \qquad Da'(+\infty) = \gamma.$$

Similarly,

$$(3.36) \qquad\qquad b(+\infty) = 0, \qquad Db'(-\infty) = -\gamma.$$

Equations (3.33), (3.35) and (3.36) constitute a boundary value problem for $a(x)$ and $b(x)$. Figure 3.4 presents a numerical solution of the nondimensionalized boundary value problem (3.43)–(3.45), to be derived shortly. The first panel shows the graphs of $a(x)$ and $b(x)$, and the second panel is a graph of the product $a(x)b(x)$, which is proportional to the rate of creation of C.

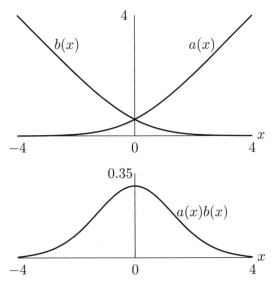

Figure 3.4.

The boundary value problem is hard and its explicit solution unknown. But basic *physical* information follows from dimensional analysis. For instance, how does the flame thickness (the length of the interval where the source kab is concentrated) depend on the influx γ?

There are three physical parameters, with units displayed in the following table:

	Parameter	D	γ	k
(3.37)				
	Unit	$\dfrac{L^2}{T}$	$\dfrac{1}{L^2 T}$	$\dfrac{L^3}{T}$

D is diffusion, so its units are $\frac{L^2}{T}$. γ is influx of A and B from $x = +\infty$ and $-\infty$; flux is a rate of particles crossing per unit area, hence its units are $\frac{1}{L^2 T}$. The unit of k follows by a balance of units in the ODEs (3.33). Let $[k]$ denote the units of k. We have

$$Da'' = kab,$$

$$\text{units} \quad \left(\frac{L^2}{T}\right)\left(\frac{1}{L^3}\right)\left(\frac{1}{L^2}\right) = [k]\left(\frac{1}{L^3}\right)^2,$$

from which it follows that $[k] = \frac{L^3}{T}$ as in (3.37).

There is a dimensionless multiplicative combination of D, γ, k. Seek it as

(3.38) $$\mu = D^a \gamma^b k^c.$$

The exponents a, b, c are restricted by balance of units. We have

$$1 = \left(\frac{L^2}{T}\right)^a \left(\frac{1}{L^2 T}\right)^b \left(\frac{L^3}{T}\right)^c = L^{2a-2b+3c} T^{-a-b-c},$$

so

$$2a - 2b + 3c = 0, \qquad a + b + c = 0,$$

and it follows from these that

$$\begin{bmatrix} a \\ b \\ c \end{bmatrix} = \begin{bmatrix} -5 \\ 1 \\ 4 \end{bmatrix} b.$$

Hence, (3.38) becomes

(3.39) $$\mu = \left(\frac{\gamma k^4}{D^5}\right)^b.$$

The essential fact is the combination $\frac{\gamma k^4}{D^5}$. Different choices of b make (irrelevant) redefinitions of μ, so we take $b = 1$. What is the importance of the dimensionless parameter μ?

Suppose we want to nondimensionalize the boundary value problem (3.33), (3.35) and (3.36), and we need to choose units for x and a, b. Now x is a length, so we ask, what combinations of D, γ, k make a length? A glance at (3.37) reveals $\frac{k}{D}$ as an obvious choice—but not the only one. Another simple choice is $\left(\frac{D}{\gamma}\right)^{1/4}$. The preceding two lengths are related by

$$\frac{k}{D} = \mu^{1/4}\left(\frac{D}{\gamma}\right)^{1/4}.$$

In general, given any specific length made of D, γ and k, we can multiply that length by an arbitrary function of μ and still have a length. For instance, the general length made of D, γ, k can be taken as $\frac{k}{D} f(\mu)$, where $f(\mu)$ is an arbitrary function of μ. Similarly, a and b are concentrations, so their units are $\frac{1}{L^3}$ and the general combination of D, γ, k with units $\frac{1}{L^3}$ is $\left(\frac{D}{k}\right)^3 g(\mu)$, where $g(\mu)$ is another arbitrary function. We summarize these results in a scaling table:

(3.40)

Variable	x	a, b
Unit	$\dfrac{k}{D} f(\mu)$	$\left(\dfrac{D}{k}\right)^3 g(\mu)$

The original ODE pair (3.33) nondimensionalized with these units reads

$$(3.41) \qquad \frac{1}{f^2} a'' = gab = \frac{1}{f^2} b''.$$

If we impose

$$(3.42) \qquad f^2 g = 1,$$

the ODEs (3.41) achieve parameter-free form

$$(3.43) \qquad a'' = ab = b''.$$

The second of the boundary conditions (3.35) nondimensionalizes to

$$a'(+\infty) = \mu \frac{f}{g}.$$

If we impose

$$(3.44) \qquad \mu \frac{f}{g} = 1,$$

we get parameter-free boundary conditions

$$(3.45) \qquad \begin{aligned} a(-\infty) &= 0, \quad a'(+\infty) = 1, \\ b(+\infty) &= 0, \quad b'(-\infty) = -1. \end{aligned}$$

From (3.42) and (3.44) it follows that

$$(3.46) \qquad f(\mu) = \mu^{-1/3}, \qquad g(\mu) = \mu^{2/3}.$$

With these determinations of $f(\mu)$ and $g(\mu)$, the scaling table (3.40) reduces to

(3.47)

Variable	x	a, b
Unit	$\dfrac{D^{2/3}}{k^{1/3}\gamma^{1/3}}$	$\dfrac{\gamma^{2/3}}{k^{1/3}D^{1/3}}$

Now for the main point of this dimensional analysis: The units of x, a, b in table (3.47) are the *unique* combinations of D, γ, k such that the original boundary value problem (3.33), (3.35) and (3.36) nondimensionalizes to the parameter-free boundary value problem (3.43) and (3.45). Let us assume that the dimensionless equations have a unique solution $a(x), b(x)$. The unique solution for dimensional a is then

$$(3.48) \qquad a_{\text{dim}} = \frac{\gamma^{2/3}}{k^{1/3}D^{1/3}} \, a\left(\frac{k^{1/3}\gamma^{1/3}}{D^{2/3}}x\right),$$

and similarly for b. From (3.48) we see that the flame thickness is the unit of x in table (3.47). Hence, the flame thickness scales with γ like $\gamma^{-1/3}$. The source kab scales with γ like $\gamma^{4/3}$. Notice that the product (flame thickness) $\times$ (source intensity) scales like γ, consistent with $\int_{-\infty}^{\infty} kab \, dx = \gamma$ in (3.34).

Problem 3.6 (Absorbing boundary condition for diffusion).   Brownian particles diffuse in $\mathbb{R}^3$ with diffusivity D. A surface S which divides $\mathbb{R}^3$ into two or more disjoint, connected regions is called *absorbing* if a particle can cross it only once. Let $c(\boldsymbol{x}, t)$ denote the concentration of particles in one of the connected regions bounded by S. The concentration c *vanishes* on S. Here, we develop a plausibility argument for this zero boundary condition by analyzing the following steady state process: Particles in $x > 0$ diffuse towards the plane $x = 0$ with a steady diffusive flux $-f\boldsymbol{e}_x$, where f is a constant, independent of time. In $x < 0$, the diffusing particles are extirpated at a rate of $-\frac{1}{\tau}c(x)$ per unit volume. Here, $c(x)$ independent of t, y and z is the steady concentration field, and τ is an "extirpation" time constant. In the limit $\tau \to 0$, the plane $x = 0$ becomes purely absorbing with respect to the particles in $x > 0$.

a) Formulate the boundary value problem for the steady concentration field $c(x)$ in $-\infty < x < \infty$, and solve it. Show that $c_0 := c(0) \to 0$ as $\tau \to 0$.

b)   We examine the functional dependence of c_0 upon parameters τ, D and f from the dimensional analysis perspective: What is the most general combination of τ, D and f that has the same units as c_0? The boundary value problem for $c(x)$ has a certain symmetry with respect to the parameter f that "selects" the answer in Part a. What is this symmetry and how does it select the answer in Part a?

Solution.

a)   The concentration $c(x)$ satisfies the ODE $Dc_{xx} = 0$ in $x > 0$ and $Dc_{xx} = \frac{c}{\tau}$ in $x < 0$. The steady flux $-Dc_x = -f$ in $x > 0$ implies $c_x \equiv \frac{f}{D}$ in $x > 0$, and this is consistent with $c_{xx} = 0$ in $x > 0$. Since c and c_x are continuous across $x = 0$, we have $c(0^+) = c(0^-) = c_0$, a common value, and

(3.6-1)
$$c_x(0^-) = \frac{f}{D}.$$

Since there is absorption of particles in $x < 0$ and no additional source, we have $c(-\infty) = 0$. The solution of $Dc_{xx} = \frac{c}{\tau}$ in $x < 0$ subject to the boundary conditions $c(0^-) = c_0$ and $c(-\infty) = 0$ is

(3.6-2)
$$c(x) = c_0 e^{\frac{x}{\sqrt{D\tau}}}.$$

Substituting (3.6-2) into (3.6-1) gives $\frac{1}{\sqrt{D\tau}} c_0 = \frac{f}{D}$, or

(3.6-3)
$$c_0 = \sqrt{\frac{\tau}{D}} f.$$

We see that c_0 vanishes like $\sqrt{\tau}$ as $\tau \to 0$.

b) First, construct a "units table":

Parameter	τ	D	f	c_0
Unit	T	$L^2 T^{-1}$	$L^{-2} T^{-1}$	L^{-3}

Here, T stands for "time" and L for "length". We propose

(3.6-4)
$$c_0 = \tau^\alpha D^\beta f^\gamma,$$

where equality means balance of units. Substituting for c_0, τ, D and f from the units table, (3.6-4) becomes

$$L^{-3} = (T)^\alpha (L^2 T^{-1})^\beta (L^{-2} T^{-1})^\gamma = L^{2(\beta-\gamma)} T^{\alpha-\beta-\gamma},$$

so $\alpha - \beta - \gamma = 0$ and $\beta - \gamma = -\frac{3}{2}$, or

(3.6-5)
$$\begin{bmatrix} 1 & -1 & -1 \\ 0 & 1 & -1 \end{bmatrix} \begin{bmatrix} \alpha \\ \beta \\ \gamma \end{bmatrix} = \begin{bmatrix} 0 \\ -\frac{3}{2} \end{bmatrix}.$$

The matrix has null vector $\begin{bmatrix} 2 \\ 1 \\ 1 \end{bmatrix}$, indicating the dimensionless combination $\tau^2 D f$. (3.6-5) has a particular solution $\begin{bmatrix} \alpha \\ \beta \\ \gamma \end{bmatrix} = \begin{bmatrix} -\frac{3}{2} \\ -\frac{3}{2} \\ 0 \end{bmatrix}$, which means that $(\tau D)^{-3/2}$ has units of L^{-3}. Hence, the general "dimensional consistency" relation is

(3.6-6)
$$c_0 = \frac{1}{(\tau D)^{\frac{3}{2}}} g(\tau^2 D f),$$

where $g(\cdot)$ is an arbitrary function. The solution of the boundary value problem for $c(x)$ is *linear* in f. Hence, c_0 should be proportional to f. It follows that $g(\cdot)$ in (3.6-6) is a linear function, and in this case, c_0 is proportional to $\sqrt{\frac{\tau}{D}} f$, as in (3.6-3).

Problem 3.7 (An "evaporation" interface driven by chemical reaction).

The liquid phase of a substance A with volume V per monomer occupies a half space $x < X(t)$ of $\mathbb{R}^3$, and there is a gas in $x > X(t)$ made mostly of monomers of another substance B, as well as some A's with concentration $c(x, t)$. The A's in the gas are dilute, i.e., $c(x, t) \ll \frac{1}{V}$. The transport of A-monomers in the gas is convection-diffusion. In particular, the velocity field is just $v e_x$, where $v = \dot{X}(t)$ is the interface velocity. A local thermodynamic equilibrium between the liquid and the gas fixes the value of c at the interface at the *solubility* value c_s, where $c_s \ll \frac{1}{V}$. In addition, A-monomers react with B's to form an inert product which cannot be reabsorbed into the A-liquid. The supply of B is sufficiently large that the reaction with A's does not significantly deplete it, so we approximate the rate of absorption of A's per unit volume by $-\frac{1}{\tau}c(x, t)$, where τ is a positive time constant.

a) Formulate the free boundary problem which governs the motion of the interface $x = X(t)$ and the evolution of $c(x, t)$ in $x > X(t)$.

b) What is the scaling unit $[x]$ of length that results from the balance between diffusive transport of A and its absorption by reaction with B? The natural scaling unit of c is of course $[c] = c_s$. What is the characteristic value $[v]$ of the interface velocity $v := \dot{X}$? What is the characteristic time $[t]$ for the interface to move a distance on the order of $[x]$? Nondimensionalize the free boundary problem for $X(t)$ and $c(x, t)$ using the above scaling units.

c) There are solutions of the free boundary problem with a steady interface velocity $\dot{X} = v = $ constant. Determine the nondimensional steady velocity from the appropriate solution of the dimensionless equations in Part b, and then "dimensionalize" it.

Solution.

a) The convection-diffusion of A's and their absorption by reaction with B's is represented by the PDE for c,

$$(3.7\text{-}1) \qquad\qquad c_t + v c_x = D c_{xx} - \frac{c}{\tau}$$

in $x > X(t)$. The interfacial boundary conditions are

(3.7-2)
$$c = c_s, \quad Dc_x = \frac{v}{V}$$

on $x = X(t)$. No A's can make it to $x = +\infty$ without being absorbed, so

(3.7-3)
$$c \to 0$$

as $x \to +\infty$.

b) The length constant associated with the decay of c away from the interface is $[x] = \sqrt{D\tau}$. In the "conservation" boundary condition $Dc_x = \frac{v}{V}$ on $x = X(t)$, the left-hand side has order of magnitude $\frac{Dc_s}{[x]} = c_s\sqrt{\frac{D}{\tau}}$, so the scaling unit of v is $[v] = Vc_s\sqrt{\frac{D}{\tau}}$. The unit of t which follows from $\frac{[x]}{[t]} = [v]$ is $[t] = \frac{\tau}{Vc_s}$. The dimensionless version of the PDE (3.7-1) is

$$\left(\frac{1}{[t]}\partial_t\right)c + [v]v\left(\frac{1}{[x]}\partial_x\right)c = D\left(\frac{1}{[x]^2}\partial_{xx}\right)c - \frac{c}{\tau},$$

or

(3.7-4)
$$c_{xx} - c = \varepsilon(c_t + vc_x)$$

in $x > X(t)$. Here, $\varepsilon := Vc_s$, which is small since $c_s \ll \frac{1}{V}$. The dimensionless boundary conditions on the interface are

(3.7-5)
$$c = 1, \quad c_x = v$$

on $x = X(t)$.

c) Assuming v to be constant in time, we seek c in the form $c = c(\zeta := x - vt)$ in $\zeta > 0$. The concentration $c(\zeta)$ satisfies the ODE boundary value problem

$$c''(\zeta) - c(\zeta) = 0 \quad \text{in } \zeta > 0,$$
$$c(0) = 1, \quad c(\infty) = 0.$$

The solution is $c(\zeta) = e^{-\zeta}$ in $\zeta > 0$. The second boundary condition of (3.7-5) now gives $v = -1$ as the dimensionless velocity. The dimensional velocity is $v = -Vc_s\sqrt{\frac{D}{\tau}}$.

Guide to bibliography. Recommended references for this chapter are Barenblatt [1], Batchelor [2], Chorin & Hald [5], Evans [11], Feynman volume I [12], and Wax [23].

The references [4, 11, 12, 23] provide background for the statistical mechanical discussion of the Smoluchowski PDE. In particular, Chorin & Hald [5] and Feynman volume I [12] give the classical argument for the relationship (3.7) between drag and diffusion coefficients. Chorin & Hald [5] introduce the entropy functional, similar to (3.14) here. In the "locality" example, we constructed the electric potential due to dipoles in a plane. Evans [11] presents a bigger picture of analogous constructions in potential theory. Batchelor [2] discusses briefly and simply the modification of local conservation laws due to the presence of sources. Barenblatt [1] is also included as background for this chapter, since there one can find many examples on dimensional analysis and similarity. The discussion of the diffusion flame is a prime example of "physical similarity". Problem 3.5 on diffusion in the Coulomb potential is another example of "intermediate asymptotics".

Part 2

Superposition

Chapter 4

Superposition of point source solutions

We transmit the essential ideas through a sequence of prototype examples. In our first examples, we have a dilute solution of particles which diffuse by Brownian motion, and they spontaneously decay into some product, with e-folding life τ. The concentration field $c(\boldsymbol{x}, t)$ satisfies

$$(4.1) \qquad\qquad c_t = D\Delta c - \frac{c}{\tau}.$$

Equation (4.1) is not a local conservation law due to the decay source term on the right-hand side; but there is no disappearance of matter, just the disappearance of the original species of particle.

One space dimension. Suppose particles are continuously introduced into $\mathbb{R}^3$ from the yz plane at a rate of s particles per unit area (of the yz plane) per unit time. A concrete realization is as follows: The particles might be products of burning in a flame layer about $x = 0$, whose thickness is smaller than we care to resolve. Particles diffuse away from $x = 0$ and eventually decay. The particles which decay are replaced by newcomers. We would expect that as time goes on, we get a concentration field $c = c(x)$ independent of y, z and time t. This $c(x)$ satisfies an ODE boundary value problem: Using $\sqrt{D\tau}$ as the unit of length and $s\sqrt{\tau/D}$ as the unit of c, this boundary value problem has nondimensional form

(4.2) $$c_{xx} - c = 0 \quad \text{in } |x| > 0,$$

(4.3) $$[c] := c(0^+) - c(0^-) = 0,$$

(4.4) $$[-c_x] := -c_x(0^+) + c_x(0^-) = 1,$$

(4.5) $$c \to 0 \quad \text{as } |x| \to \infty.$$

The concentration c is continuous at $x = 0$ as in (4.3) because a jump discontinuity induces a huge diffusive flux concentrated near $x = 0$. There would be an immediate redistribution of particles which erases the jump in c. In (4.4), $-c_x(0^+)$ is the flux of particles into the right half space $x > 0$, and $c_x(0^-)$ is the flux into the left half space $x < 0$. Hence, $[-c_x]$ in (4.4) is the total flux from the origin, nondimensionalized to 1. The solution of the boundary value problem (4.2)–(4.5) is elementary:

(4.6) $$c = \frac{e^{-|x|}}{2}.$$

If particles are introduced at dimensionless rate per unit area s at $x = x'$, we have by homogeneity and linearity that

(4.7) $$c = s\,\frac{e^{-|x-x'|}}{2}.$$

Now suppose that the layer in which particles are introduced has finite thickness; then the point source boundary value problem (4.2)–(4.5) is replaced by

(4.8) $$c_{xx} - c = -s(x) \quad \text{in } -\infty < x < \infty,$$

(4.9) $$c \to 0 \quad \text{as } |x| \to \infty.$$

Here, the dimensionless source $s(x)$ is a function which is continuous and integrable on $-\infty < x < \infty$. The boundary value problem (4.8)–(4.9) is solved by the elementary variation of parameters method. The solution is

(4.10) $$c(x) = \int_{-\infty}^{\infty} s(x')\,\frac{e^{-|x-x'|}}{2}\,dx'.$$

The interpretation of (4.10) as a *superposition* of sheet source solutions (4.7) is clear: The total rate at which particles appear in $(x', x' + dx')$ per unit area of yz plane is $s(x')\,dx'$, so this layer acts like a sheet source of strength $s(x')\,dx'$, and the corresponding contribution to $c(x)$ is

$$s(x')\,dx'\,\frac{e^{-|x-x'|}}{2}.$$

The full solution due to all the intervals $(x', x' + dx')$ is the integral (4.10).

Problem 4.1 (Soviet rod factory). [1] $N \gg 1$ long rods slide along "runners" parallel to the x-axis. They are supposed to travel at uniform velocity $u > 0$, but in Soviet factory, their actual motion is combined uniform motion and Brownian motion: For each rod, let $y(t)$ be the displacement of a material point on the rod, relative to uniform motion at velocity u. Assuming $y(0) = 0$, the distribution of $y(t)$ at time $t > 0$ is the diffusion kernel with diffusion coefficient D. In addition, there is a "cutter" at the origin $x = 0$ of each runner, who chops the passing rod into segments. There is supposed to be a precise time interval τ between chops, but in Soviet factory, the actual time sequence of chops is random, and the best you can say is that the expected number of chops in a time interval of duration t is $\frac{t}{\tau}$. Notice that due to the random component of rod motion, the cut end of a rod may be in $x < 0$, and the next chop falls on an empty runner.

a) For each rod, let $x(t)$ be the position of its cut end. When the factory is in steady operation, the density $\rho = \rho(x)$ of these $x(t)$ is time-independent. Derive a boundary value problem for $\rho(x)$, and solve it. Explain in quantitative detail the jump discontinuity of $\rho'(x)$ at $x = 0$.

b)   As this illustrious factory operates, an inventory of rod segments is built up. What is the distribution of lengths? What is the average rate of production of rods, of all lengths? Why is the rate of production much less than the "nominal" value $\frac{N}{\tau}$ as $\frac{u^2\tau}{D} \to 0$?

Solution.

a) The transport of $x(t)$'s is convection-diffusion, with velocity u and diffusion coefficient D. The rods with $x > 0$ are subject to cutting: The rate at which rods with their ends between $x > 0$ and $x + dx > x$ are cut is $\frac{\rho(x)}{\tau} \, dx$. Rods with their ends in $x < 0$ are of course not subject to cutting. Hence, we surmise that $\rho(x)$ satisfies the ODE

$$u\rho' - D\rho'' = \left[\begin{array}{ll} 0, & x < 0, \\[2mm] -\frac{\rho}{\tau}, & x > 0. \end{array} \right.$$

At $x = 0$, ρ is continuous, and $\rho \to 0$ as $|x| \to \infty$.

Given $\rho(0)$, the solution for $\rho(x)$ is

$$\rho(x) = \left[\begin{array}{ll} \rho(0)e^{\frac{ux}{D}}, & x < 0, \\[2mm] \rho(0)e^{-kx}, & x > 0, \end{array} \right.$$

---

[1] Inspired by problems in Chernyak & Rose [**3**].

where

$$k := \frac{u}{2D}\left\{\sqrt{1 + \frac{4D}{u^2\tau}} - 1\right\}.$$

We can determine $\rho(0)$ from the normalization condition $\int_{-\infty}^{\infty} \rho(x)\,dx = N$. We find that

(4.1-1) $$\rho(0) = \frac{4N}{u\tau}\frac{1}{\left(\sqrt{1 + \frac{4D}{u^2\tau}} + 1\right)^2}.$$

The jump in ρ' at $x = 0$ is

$$[\rho'] = -\rho(0)\left\{k + \frac{u}{D}\right\} = -\frac{1}{2}\rho(0)\frac{u}{D}\left\{\sqrt{1 + \frac{4D}{u^2\tau}} + 1\right\},$$

indicating that $x = 0$ is a *source* of $x(t)$'s, of strength

(4.1-2) $$S := -D[\rho'] = \frac{1}{2}\rho(0)u\left\{\sqrt{1 + \frac{4D}{u^2\tau}} + 1\right\}.$$

Why *is* $x = 0$ a source? When an $x > 0$ rod is cut, its endpoint is "reset" to $x = 0$, so the rate at which $x(t)$'s are "emitted" from $x = 0$ must be the total cutting rate

$$\frac{1}{\tau}\int_0^{\infty} \rho\,dx = \frac{\rho(0)}{\tau k} = \frac{\rho(0)}{\tau}\frac{2D}{u}\frac{1}{\sqrt{1 + \frac{4D}{u^2\tau}} - 1} = \frac{1}{2}\rho(0)u\left\{\sqrt{1 + \frac{4D}{u^2\tau}} + 1\right\}.$$

The right-hand side is indeed the source strength S.

b) Segments with lengths between $x > 0$ and $x + dx > x$ are produced at rate $\frac{\rho(x)}{\tau}\,dx = \frac{\rho(0)}{\tau}e^{-kx}\,dx$, so the length distribution is ke^{-kx}. The total rate of production is S in (4.1-2). Using (4.1-1) to substitute for $\rho(0)$, we get

$$S = \frac{N}{\tau}\frac{2}{\sqrt{1 + \frac{4D}{u^2\tau}} + 1}.$$

In the limit $\frac{D}{u^2\tau} \to 0$, we get the "nominal" rate $\frac{N}{\tau}$. Since random motion is small in this limit, the cut end most likely moves to the right and the next strike of the cutter most likely results in another cut. If $\frac{D}{u^2\tau} \gg 1$, then most of the cut ends are in $x < 0$, and most of the strikes are just hitting empty runners.

Three space dimensions. The $\mathbb{R}^3$ version of the boundary value problem (4.8)–(4.9),

(4.11)
$$\Delta c - c = -s(\boldsymbol{x}), \quad \boldsymbol{x} \text{ in } \mathbb{R}^3,$$

(4.12)
$$c \to 0 \quad \text{as } |\boldsymbol{x}| \to \infty,$$

falls to a similar analysis. First, suppose that particles are introduced at dimensionless rate s at a single point $\boldsymbol{x} = \boldsymbol{x}'$. The concentration field satisfies the homogeneous PDE with $s \equiv 0$ at all $\boldsymbol{x} \neq \boldsymbol{x}'$. The net outflux of particles from $\boldsymbol{x} = \boldsymbol{x}'$ is expressed by

(4.13)
$$s := \lim_{\varepsilon \to 0} \int_{|\boldsymbol{x}-\boldsymbol{x}'|=\varepsilon} -\nabla c \cdot \boldsymbol{e}_r \, da.$$

The boundary value problem consisting of PDE (4.11) with $s \equiv 0$ in $\boldsymbol{x} \neq \boldsymbol{x}'$, the zero boundary condition (4.12) at ∞, and the flux condition (4.13) is invariant under rotation about any axis through $\boldsymbol{x} = \boldsymbol{x}'$, so one expects a spherically symmetric solution $c = c(r := |\boldsymbol{x} - \boldsymbol{x}'|)$. The radially symmetric boundary value problem is:

(4.14)
$$c_{rr} + \frac{2}{r} c_r - c = 0 \quad \text{in } r > 0,$$
$$c(\infty) = 0,$$
$$s = -\lim_{\varepsilon \to 0} 4\pi\varepsilon^2 c_r(\varepsilon).$$

Linear PDEs with the three-dimensional radial Laplacian $c_{rr} + \frac{2}{r}c_r$ are conquered by a nice trick: We have

$$c_{rr} + \frac{2}{r}c_r = \frac{1}{r}(rc)_{rr},$$

so the ODE in (4.14) becomes

$$(rc)_{rr} - (rc) = 0.$$

The general solution for rc is a linear combination of e^{+r} and e^{-r}. The zero boundary condition at ∞ selects the decaying exponential, and we have

$$c = \frac{C}{r} e^{-r},$$

where C is a constant. The constant is determined by the flux condition, which is the last line of (4.14),

$$s = -\lim_{\varepsilon \to 0} 4\pi\varepsilon^2 c_r(\varepsilon) = \lim_{\varepsilon \to 0} 4\pi C(1+\varepsilon)e^{-\varepsilon} = 4\pi C.$$

So $C = \frac{s}{4\pi}$ and finally,

(4.15)
$$c(r) = s\frac{e^{-r}}{4\pi r} = s\frac{e^{-|\boldsymbol{x}-\boldsymbol{x}'|}}{4\pi|\boldsymbol{x}-\boldsymbol{x}'|}.$$

A basic difference between (4.15) and its $\mathbb{R}^1$ counterpart (4.7) is the geometric factor $\frac{1}{|\boldsymbol{x}-\boldsymbol{x}'|}$. Under steady conditions, the diffusive outflux through a sphere of radius ε about $\boldsymbol{x}'$,

$$-\int_{|\boldsymbol{x}-\boldsymbol{x}'|=\varepsilon} \nabla c \cdot \boldsymbol{e}_r \, da,$$

is asymptotically constant as $\varepsilon \to 0$. For this to happen, we need ∇c to be proportional to $\frac{1}{r^2}$, hence the factor $\frac{1}{r}$ in c.

The heuristic construction of the solution to (4.11)–(4.12) with a continuous source $s(\boldsymbol{x})$ is a superposition analogous to (4.10) for the $\mathbb{R}^1$ case: Particles are introduced into a volume element $d\boldsymbol{x}'$ about $\boldsymbol{x} = \boldsymbol{x}'$ at the rate $s(\boldsymbol{x}') \, d\boldsymbol{x}'$, and the corresponding contribution to $c(\boldsymbol{x})$ is (4.15) with $s(\boldsymbol{x}') \, d\boldsymbol{x}'$ in place of S. That is,

$$\frac{s(\boldsymbol{x}') \, d\boldsymbol{x}'}{4\pi} \frac{e^{-|\boldsymbol{x}-\boldsymbol{x}'|}}{|\boldsymbol{x} - \boldsymbol{x}'|}.$$

The full solution obtained by summing over all volume elements $d\boldsymbol{x}'$ is

$$(4.16) \qquad c(\boldsymbol{x}) = \int_{\mathbb{R}^3} S(\boldsymbol{x}') \frac{e^{-|\boldsymbol{x}-\boldsymbol{x}'|}}{4\pi|\boldsymbol{x} - \boldsymbol{x}'|} \, d\boldsymbol{x}'.$$

This integral is improper due to the singularity at $\boldsymbol{x}' = \boldsymbol{x}$, but its convergence is strong. Let us write the integral (4.16) in geometric form. First, write $\boldsymbol{x}' = \boldsymbol{x} + r\boldsymbol{e}_r$, where r and $\boldsymbol{e}_r$ are the radial coordinate and radial unit vector about $\boldsymbol{x}' = \boldsymbol{x}$. The volume element becomes $r^2 \, dr \, d\Omega$, where $d\Omega$ denotes the area element on the unit sphere S^2. If we introduce the usual angles ϑ and φ as coordinates of S^2, we get $d\Omega = \sin\vartheta \, d\vartheta \, d\varphi$. The integral (4.16) becomes

$$c(\boldsymbol{x}) = \int_0^\infty \int_{\boldsymbol{e}_r \text{ on } S^2} s(\boldsymbol{x} + r\boldsymbol{e}_r) \left(\frac{e^{-r}}{4\pi r}\right) (r^2 \, dr \, d\Omega),$$

and upon rearranging we get

$$(4.17) \qquad c(\boldsymbol{x}) = \int_0^\infty \left(\frac{1}{4\pi} \int_{\boldsymbol{e}_r \text{ on } S^2} s(\boldsymbol{x} + r\boldsymbol{e}_r) \, d\Omega\right) r e^{-r} \, dr.$$

The term in parentheses is the average of $s(\boldsymbol{x}')$ over the sphere of radius r about $\boldsymbol{x}' = \boldsymbol{x}$. If $s(\cdot)$ is continuous, then the average converges to $s(\boldsymbol{x})$ as $r \to 0$, and the integrand in (4.17) is not only integrable at $r = 0$ but in fact vanishes there.

The construction of solution (4.16) by superposition of point sources is heuristic. One can (reasonably) ask for an authentic proof. We will do proof by Green's identity as an exercise. That proof uses the point source solution (4.15) as an *ingredient*.

Problem 4.2 (Superposition solution by Green's identity).

a) Let $c(\boldsymbol{x})$ and $g(\boldsymbol{x})$ be analytic functions of $\boldsymbol{x}$ in a region R of $\mathbb{R}^3$. Derive *Green's identity*

$$(4.2\text{-}1) \qquad \int_R (c\Delta g - g\Delta c)(\boldsymbol{x})\, dx = \int_{\partial R} (cg_n - gc_n)\, da.$$

b) Let $s(\boldsymbol{x})$ be a function of $\boldsymbol{x}$ in $\mathbb{R}^3$ whose support is bounded. In Green's identity (4.2-1), take $c(\boldsymbol{x})$ to be the solution of

$$(4.2\text{-}2) \qquad (\Delta c)(\boldsymbol{x}) - c(\boldsymbol{x}) = -s(\boldsymbol{x})$$

which vanishes as $|x| \to \infty$, and for *fixed* $\boldsymbol{X}$ in $\mathbb{R}^3$, take

$$(4.2\text{-}3) \qquad g(\boldsymbol{x}) = \frac{e^{-|\boldsymbol{X}-\boldsymbol{x}|}}{4\pi|\boldsymbol{X}-\boldsymbol{x}|}.$$

The region R is the "shell" of $\boldsymbol{x}$ with $\varepsilon < |\boldsymbol{x}-\boldsymbol{X}| < \rho$. Evaluate the resulting integral identity in the limit $\varepsilon \to 0$, $\rho \to \infty$.

Solution.

a) Integrating the "product rule identity"

$$c\Delta g - g\Delta c = \boldsymbol{\nabla} \cdot (c\boldsymbol{\nabla}g - g\boldsymbol{\nabla}c)$$

over R and using the divergence theorem on the right-hand side gives

$$\int_R (c\Delta g - g\Delta c)(\boldsymbol{x})\, dx' = \int_{\partial R} (cg_n - gc_n)\, da.$$

b) By (4.2-2) we have $(\Delta c)(\boldsymbol{x}) = c(\boldsymbol{x}) - s(\boldsymbol{x})$ in R, and $(\Delta g)(\boldsymbol{x}) = g(\boldsymbol{x})$ in R from (4.2-3). Doing these substitutions for Δc and Δg, the left-hand side of Green's identity (4.2-1) becomes

$$\int_R g(\boldsymbol{x})s(\boldsymbol{x})\, dx = \int_{\varepsilon < |\boldsymbol{x}-\boldsymbol{X}| < \rho} s(\boldsymbol{x}) \frac{e^{-|\boldsymbol{X}-\boldsymbol{x}|}}{4\pi|\boldsymbol{X}-\boldsymbol{x}|}\, dx.$$

The integral converges to

$$\int_{\mathbb{R}^3} s(\boldsymbol{x}) \frac{e^{-|\boldsymbol{X}-\boldsymbol{x}|}}{4\pi|\boldsymbol{X}-\boldsymbol{x}|}\, dx$$

as $\varepsilon \to 0$, $\rho \to \infty$. Now look at the surface integral on the right-hand side of (4.2-1). The contribution from $|\boldsymbol{x}-\boldsymbol{X}| = \rho$ vanishes as $\rho \to \infty$, and the contribution from $|\boldsymbol{x}-\boldsymbol{X}| = \varepsilon$ is

$$\frac{1}{4\pi} \int_{\boldsymbol{e}_r \text{ on } S^2} \left\{ c(\boldsymbol{X} + \varepsilon\boldsymbol{e}_r)\left(\frac{1}{\varepsilon^2} + \frac{1}{\varepsilon}\right) - (\boldsymbol{\nabla}c)(\boldsymbol{X} + \varepsilon\boldsymbol{e}_r) \cdot \boldsymbol{e}_r \left(\frac{1}{\varepsilon}\right) \right\} e^{-\varepsilon}\varepsilon^2\, d\Omega.$$

Here, $d\Omega$ is the area element on the unit sphere S^2, as in (4.17). The limit as $\varepsilon \to 0$ is simply $c(\boldsymbol{X})$. Hence, the $\varepsilon \to 0$, $\rho \to \infty$ limit of Green's identity is

$$c(\boldsymbol{X}) = \int_{\mathbb{R}^3} c(\boldsymbol{x}') \frac{e^{-|\boldsymbol{X}-\boldsymbol{x}|}}{4\pi|\boldsymbol{X}-\boldsymbol{x}|}\, d\boldsymbol{x}.$$

Problem 4.3 (The smoke plume).   Particles in $\mathbb{R}^3$ minus the origin undergo convection-diffusion. The velocity field is $u\boldsymbol{e}_1$, with u a positive constant, and the diffusion coefficient is $D > 0$. Particles emerge out of the origin $\boldsymbol{x} = 0$ at the rate of S particles per unit time.

a) Formulate the *time-independent* convection-diffusion PDE for the concentration field $c(\boldsymbol{x})$. What is the flux vector field $\boldsymbol{f}$? What is the surface integral of $\boldsymbol{f}$ over any surface enclosing the origin?

b) Determine the solutions $c(\boldsymbol{x})$ of the PDE which vanish as $|\boldsymbol{x}| \to \infty$ and for which $\lim_{\boldsymbol{x} \to 0} |\boldsymbol{x}|c(\boldsymbol{x})$ exists. (Hint: Take $c(\boldsymbol{x}) = e^{k x_1} g(\boldsymbol{x})$, with k constant, and formulate the PDE for $g(\boldsymbol{x})$. There is a special choice of k which *removes* the first derivative term from the PDE for $g(\boldsymbol{x})$. What is left should look familiar.)

c) The solutions in Part b contain a proportionality constant. Determine this constant so that the solution is consistent with a source of strength S at the origin. (Hint: $\int_{r=\varepsilon} \boldsymbol{f} \cdot \boldsymbol{e}_r\, da$.)

d) Show that the solution $c(\boldsymbol{x})$ satisfies the downstream flux condition

(4.3-1) $$\int_{\mathbb{R}^2} (uc - D\partial_1 c)(X, x_2, x_3)\, dx_2 dx_3 = S$$

for any $X > 0$.

Solution.

a) The PDE is

(4.3-2) $$\boldsymbol{\nabla} \cdot \boldsymbol{f} = 0,$$

where $\boldsymbol{f}$ is the flux

(4.3-3) $$\boldsymbol{f} = cu\boldsymbol{e}_1 - D\boldsymbol{\nabla} c.$$

It follows from (4.3-2) and (4.3-3) that

(4.3-4) $$u\partial_1 c - D\Delta c = 0$$

in $\mathbb{R}^3$ minus the origin. For any region R with the source $\boldsymbol{x} = 0$ in its interior,

$$(4.3\text{-}5) \qquad \int_{\partial R} \boldsymbol{f} \cdot \boldsymbol{n} \, da = S.$$

b) From $c(\boldsymbol{x}) = e^{kx_1} g(\boldsymbol{x})$, we calculate

$$\boldsymbol{\nabla} c = e^{kx_1}(\boldsymbol{\nabla} g + kg\boldsymbol{e}_1),$$
$$\partial_1 c = \boldsymbol{\nabla} c \cdot \boldsymbol{e}_1 = e^{kx_1}(\partial_1 g + kg),$$
$$\Delta c = \boldsymbol{\nabla} \cdot \boldsymbol{\nabla} c = e^{kx_1}\{\Delta g + k\partial_1 g + k\boldsymbol{e}_1 \cdot (\boldsymbol{\nabla} g + kg\boldsymbol{e}_1)\}$$
$$= e^{kx_1}\{\Delta g + 2k\partial_1 g + k^2 g\}.$$

Hence, (4.3-4) transforms into the PDE for g,

$$u(\partial_1 g + kg) - D(\Delta g + 2k\partial_1 g + k^2 g) = 0,$$

and upon choosing $k = \frac{u}{2D}$, this reduces to

$$\Delta g - k^2 g = 0.$$

The spherically symmetric solutions for g which vanish as $r \to \infty$ are

$$g = C\frac{e^{-kr}}{r},$$

where C is an arbitrary constant. The corresponding solutions for c are

$$(4.3\text{-}6) \qquad c = C\frac{e^{k(x_1-r)}}{r}.$$

It is clear that these vanish as $r \to \infty$ and that

$$\lim_{\boldsymbol{x} \to 0} rc(\boldsymbol{x}) = C.$$

Notice that the decay of c is *exponential* along any ray not parallel to the positive x_1-axis, and algebraic like $\frac{1}{r}$ along the positive x_1-axis. The positive x_1-axis is "downstream" from the source $\boldsymbol{x} = 0$, so most of the smoke should be concentrated about the positive x_1-axis.

c) It follows from (4.3-6) that

$$\boldsymbol{\nabla} c = C\left\{-\frac{\boldsymbol{e}_r}{r^2} + \frac{k}{r}(\boldsymbol{e}_1 - \boldsymbol{e}_r)\right\} e^{k(x_1-r)} = -\frac{C}{r^2}\boldsymbol{e}_r + O\left(\frac{1}{r}\right)$$

as $r \to 0$, and that the flux in (4.3-3) is

$$\boldsymbol{f} = \frac{DC}{r^2}\boldsymbol{e}_r + O\left(\frac{1}{r}\right)$$

as $r \to 0$. In (4.3-5) take the region R to be the ball of radius ε about $\boldsymbol{x} = 0$.
We have

$$S = \int_{\partial R} \boldsymbol{f} \cdot \boldsymbol{n} \, da = 4\pi\varepsilon^2 \left(\frac{DC}{\varepsilon^2} + O\left(\frac{1}{\varepsilon}\right) \right) \to 4\pi DC$$

as $\varepsilon \to 0$. Hence, $C = \frac{S}{4\pi D}$, and (4.3-6) becomes

(4.3-7) $$c(\boldsymbol{x}) = \frac{S}{4\pi D} \frac{e^{\frac{u}{2D}(x_1 - r)}}{r}.$$

Here, we used $k = \frac{u}{2D}$.

d) Let R be the half ball $x_1 < X$, $|\boldsymbol{x} - X\boldsymbol{e}_1| < \rho$, with $X > 0$. For $\rho > X$,
the origin $\boldsymbol{x} = 0$ is inside R, and

$$\int_{\partial R} \boldsymbol{f} \cdot \boldsymbol{n} \, da = S.$$

In the limit $\rho \to \infty$, the contribution from the hemisphere $|\boldsymbol{x} - X\boldsymbol{e}_1| = \rho$,
$x_1 < X$, vanishes, and what is left is

$$\int_{\mathbb{R}^2} \boldsymbol{f}(X, x_2, x_3) \cdot \boldsymbol{e}_1 \, dx_2 dx_3 = S.$$

Substituting (4.3-3) for $\boldsymbol{f}$ gives (4.3-1).

Problem 4.4 (Far downstream limit).   We examine the plume solution
(4.3-7) in the far downstream limit. The solution contains the length con-
stant $\frac{D}{u}$, and we introduce the scaling unit of x, $[x] := \frac{D}{u}\frac{1}{\varepsilon}$, where $\varepsilon > 0$ is
a gauge parameter; the limit $\varepsilon \to 0$ is considered.

a) Find the appropriate scaling unit of x_2 and x_3, based upon the thickness
of the plume at downstream distances on the order of $[x]$. What is the
corresponding scaling unit $[c]$ of the concentration field?

b) Nondimensionalize the solution (4.3-7) using the scaling units of the variables developed in Part a, and take its $\varepsilon \to 0$ limit. Explain the connection
to the $\mathbb{R}^2$ diffusion kernel.

c) Nondimensionalize the time-independent convection-diffusion equation in
Problem 4.3a and the "downstream flux condition" (4.3-1) using the scaling
units in Part a. Take the limits $\varepsilon \to 0$. What is the solution of the $\varepsilon \to 0$
equations which vanishes as $x_1 \to 0$ with $\sqrt{x_2^2 + x_3^2} > 0$ fixed?

Solution.

a) A particle is convected downstream a distance $[x]$ in time $\frac{[x]}{u}$, and the lateral diffusion in the x_2 or x_3 directions in this time is on the order of $\sqrt{\frac{D[x]}{u}} = \frac{D}{u}\frac{1}{\sqrt{\varepsilon}}$. Hence, the scaling unit of x_2 and x_3 is $\frac{D}{u}\frac{1}{\sqrt{\varepsilon}}$. The unit of c discerned from the solution (4.3-7) is

(4.4-1) $$[c] = \frac{S}{D}\frac{1}{[x]} = \frac{Su}{D^2}\varepsilon.$$

A more physical argument that does *not* rely on knowledge of the solution is the following: Stuff flows past the plane $x_1 = \frac{D}{u}\frac{1}{\varepsilon}$ at velocity u, in a plume of thickness $\frac{D}{u}\frac{1}{\sqrt{\varepsilon}}$. Hence, we have an order-of-magnitude balance

(4.4-2) $$S = [c]u\left(\frac{D}{u}\frac{1}{\sqrt{\varepsilon}}\right)^2.$$

Here, $\left(\frac{D}{u}\frac{1}{\sqrt{\varepsilon}}\right)^2$ is the cross-sectional area of the plume. Solving (4.4-2) for $[c]$ recovers (4.4-1).

b) First, the dimensional r in terms of dimensionless x_1, x_2 and x_3 is

$$r = \left\{\left(\frac{D}{u\varepsilon}x_1\right)^2 + \left(\frac{D}{u\sqrt{\varepsilon}}x_2\right)^2 + \left(\frac{D}{u\sqrt{\varepsilon}}x_3\right)^2\right\}^{\frac{1}{2}}$$
$$= \frac{D}{u}\frac{1}{\varepsilon}\sqrt{x_1^2 + \varepsilon x_2^2 + \varepsilon x_3^2},$$

and the exponent $\frac{u}{4D}(x_1 - r)$ in (4.3-7) in terms of dimensionless variables is

$$\frac{1}{2\varepsilon}\left\{x_1 - \sqrt{x_1^2 + \varepsilon x_2^2 + \varepsilon x_3^2}\right\}.$$

Now we do the nondimensionalization of (4.3-7):

$$\left(\frac{Su}{D^2}\varepsilon\right)c = \frac{S}{4\pi D}\frac{e^{\frac{1}{2\varepsilon}\left\{x_1 - \sqrt{x_1^2 + \varepsilon x_2^2 + \varepsilon x_3^2}\right\}}}{\frac{D}{u\varepsilon}\sqrt{x_1^2 + \varepsilon x_2^2 + \varepsilon x_3^2}},$$

and this simplifies to

$$c = \frac{1}{4\pi}\frac{e^{\frac{1}{2\varepsilon}\left\{x_1 - \sqrt{x_1^2 + \varepsilon x_2^2 + \varepsilon x_3^2}\right\}}}{\sqrt{x_1^2 + \varepsilon x_2^2 + \varepsilon x_3^2}}.$$

The $\varepsilon \to 0$ limit is

(4.4-3) $$c^0 = \frac{1}{4\pi x_1}e^{-\frac{x_2^2 + x_3^2}{4x_1}}.$$

Let $g(r, t)$ be the $\mathbb{R}^2$ diffusion kernel with unit diffusion ($D = 1$). We recognize (4.4-3) as $g(\sqrt{x_2^2 + x_3^2}, x_1)$. Notice that x_1 is analogous to "time".

c) We nondimensionalize the PDE $u\partial_1 c - D\Delta c = 0$:

$$u\left(\frac{u}{D}\varepsilon\partial_1\right) = D\left\{\left(\frac{u}{D}\varepsilon\,\partial_1\right)^2 c + \left(\frac{u}{D}\sqrt{\varepsilon}\,\partial_2\right)^2 c + \left(\frac{u}{D}\sqrt{\varepsilon}\,\partial_3\right)^2 c\right\},$$

which simplifies to

(4.4-4) $$\partial_1 c = \varepsilon\partial_{11}c + \partial_{22}c + \partial_{33}c.$$

The nondimensionalization of the flux identity

$$S = \int_{\mathbb{R}^2} (uc - D\partial_1 c)(x_1, x_2, x_3)\, dx_2 dx_3$$

for $x_1 > 0$ is

$$S = u[c]\int_{\mathbb{R}^2}\left\{c - \frac{D}{u}\left(\frac{u}{D}\varepsilon\partial_1\right)c\right\}\left(\frac{D}{u}\frac{1}{\sqrt{\varepsilon}}\right)^2 dx_2 dx_3,$$

and using the $[c]$ in (4.4-1), this simplifies to

(4.4-5) $$1 = \int_{\mathbb{R}^2}\{c - \varepsilon\partial_1 c\}(x_1, x_2, x_3)\, dx_2 dx_3.$$

In the limit $\varepsilon \to 0$, (4.4-4) and (4.4-5) reduce to

(4.4-6) $$\partial_1 c = \partial_{22}c + \partial_{33}c,$$

(4.4-7) $$\int_{\mathbb{R}^2} c(x_1, x_2, x_3)\, dx_2 dx_3 = 1, \quad x_1 > 0.$$

Equation (4.4-6) is a diffusion PDE with $D = 1$ and x_1 as "time". The solution which satisfies the integral condition (4.4-7) and vanishes as $x_1 \to 0$ with $\sqrt{x_2^2 + x_3^2} > 0$ fixed *is* the $\mathbb{R}^2$ diffusion kernel as found in Part b.

Sources in spacetime. In the next examples, we construct concentration fields of Brownian particles due to a space- and time-dependent source $s(x, t)$, by superposition of "point source solutions in spacetime". A "point source of Brownian particles in spacetime" means a certain number of particles introduced at a *single* event (x', t'). We have seen this before in the original discussion of the diffusion kernel. If N particles are introduced at the event (x', t'), their concentration field at time $t > t'$ is $Ng(x - x', t - t')$, where g is the diffusion kernel (2.39). The concentration field generated by a continuous source $s(x, t)$ satisfies

(4.18) $$c_t - Dc_{xx} = s(x, t).$$

The construction of $c(x, t)$ by superposition of diffusion kernels is simple: The number of particles introduced in the space interval $(x', x' + dx')$ in time interval $(t', t' + dt')$ is $s(x', t') \, dx' dt'$, and their contribution to $c(x, t)$ at a time $t > t'$ is

$$\left(s(x', t') \, dx' dt' \right) g(x - x', t - t').$$

The total concentration field due to particles introduced at all events (x', t') with $t' < t$ is

(4.19)
$$
\begin{aligned}
c(x, t) &= \int_{-\infty}^{t} \int_{-\infty}^{\infty} s(x', t') g(x - x', t - t') \, dx' dt' \\
&= \int_{-\infty}^{t} \int_{-\infty}^{\infty} \frac{s(x', t')}{2\sqrt{\pi D(t - t')}} e^{-\frac{(x - x')^2}{4D(t - t')}} \, dx' dt'.
\end{aligned}
$$

There is a Green's identity calculation (Problem 5.5) which validates (4.19). Suppose the source is confined along a *curve* $x = a(t)$ in spacetime. In time dt', the number of particles released along the segment between times t' and $t' + dt'$ is $S(t') \, dt'$. Their contribution to the concentration field at time $t > t'$ is

$$\left(S(t') \, dt' \right) g(x - a(t'), t - t'),$$

and the concentration field at time t due to particles released at all times $t' < t$ is

$$c(x, t) = \int_{-\infty}^{t} S(t') g(x - a(t'), t - t') \, dt'$$

or

(4.20)
$$c(x, t) = \int_{-\infty}^{t} \frac{S(t')}{2\sqrt{\pi D(t - t')}} e^{-\frac{(x - a(t'))^2}{4D(t - t')}} \, dt'.$$

We examine this tricky looking integral representation of c, to see if the total number of particles is increasing at the correct rate. Does it really follow from (4.20) that

(4.21)
$$\frac{d}{dt} \int_{-\infty}^{\infty} c(x, t) \, dx = S(t) \, ?$$

We write

$$\int_{-\infty}^{\infty} c \, dx = \int_{-\infty}^{a(t)} c \, dx + \int_{a(t)}^{\infty} c \, dx,$$

so

$$\frac{d}{dt} \int_{-\infty}^{\infty} c \, dx = \int_{-\infty}^{a(t)} c_t \, dx + \int_{a(t)}^{\infty} c_t \, dx + \{ c(a(t)^+, t) - c(a(t)^-, t) \} \dot{a}(t).$$

The concentration c is continuous across the source, so the last term on the right-hand side vanishes. Substitute $c_t = D c_{xx}$ into the remaining integrals and use the fundamental theorem of calculus. We get

$$(4.22) \qquad \frac{d}{dt} \int_{-\infty}^{\infty} c \, dx = Dc_x(a(t)^-, t) - Dc_x(a(t)^+, t)$$

$$= -D[c_x](x = a(t), t).$$

We need to compute the limits of c_x as $x \to a(t)^+$ and as $x \to a(t)^-$ from (4.20). We introduce nondimensional variables so that $D = \frac{1}{4}$. Then (4.20) becomes

$$c(x, t) = \int_{-\infty}^{t} \frac{S(t')}{\sqrt{\pi(t - t')}} e^{-\frac{(x - a(t'))^2}{(t - t')}} \, dt'.$$

Next, change the integration variable from t' to $\tau = t - t'$:

$$c(x, t) = \int_{0}^{\infty} \frac{S(t - \tau)}{\sqrt{\pi\tau}} e^{-\frac{(x - a(t - \tau))^2}{\tau}} \, d\tau.$$

Now differentiate with respect to x and set $x = a(t) + \varepsilon$, $\varepsilon > 0$:

$$c_x(a(t) + \varepsilon, t) = -\int_{0}^{\infty} \frac{S(t - \tau)}{\sqrt{\pi\tau}} \frac{2(a(t) - a(t - \tau) + \varepsilon)}{\tau} e^{-\frac{(a(t) - a(t - \tau) + \varepsilon)^2}{\tau}} \, d\tau.$$

The exponent $\frac{(a(t) - a(t - \tau) + \varepsilon)^2}{\tau}$ suggests that the main contribution to the integral comes from $\tau = O(\varepsilon^2)$, so let us change the variable of integration again, from τ to $\sigma := \frac{\tau}{\varepsilon^2}$. We get

$$(4.23) \quad c_x(a(t) + \varepsilon, t) = -\int_{0}^{\infty} S(t - \varepsilon^2\sigma) \left\{ 1 + \frac{a(t) - a(t - \varepsilon^2\sigma)}{\varepsilon} \right\}$$

$$\times \frac{2}{\sqrt{\pi} \, \sigma^{3/2}} e^{-\left\{ 1 + \frac{a(t) - a(t - \varepsilon^2\sigma)}{\varepsilon} \right\}^2 \frac{1}{\sigma}} \, d\sigma.$$

In the limit $\varepsilon \to 0^+$,

$$\frac{a(t) - a(t - \varepsilon^2\sigma)}{\varepsilon} \sim \varepsilon\sigma\dot{a}(t) \to 0,$$

and (4.23) collapses to

$$c_x(a(t)^+, t) = -S(t) \int_{0}^{\infty} \frac{2}{\sqrt{\pi}} \frac{e^{-1/\sigma}}{\sigma^{3/2}} \, d\sigma.$$

One more substitution: Setting $s^2 = \frac{1}{\sigma}$, we get

$$c_x(a(t)^+, t) = -S(t) \int_{0}^{\infty} \frac{4}{\sqrt{\pi}} e^{-s^2} \, ds = -2S(t).$$

Similarly,

$$c_x(a(t)^-, t) = +2S(t).$$

Hence,

$$[c_x](a(t), t) = -4S(t),$$

and (4.22) with $D = \frac{1}{4}$ becomes

$$\frac{d}{dt} \int_{-\infty}^{\infty} c \, dx = -\frac{1}{4}(-4S(t)) = S(t).$$

Problem 4.5 (Moving source).   Evaluate the superposition solution (4.20) for a line source whose source strength $S(t)$ is a positive constant S and whose world line is uniform motion, with $a(t) = ut$ for some positive constant u.

Solution. With $S(t) \equiv S$ and $a(t) = ut$, (4.20) reads

(4.5-1) $$c(x, t) = \int_{-\infty}^{t} \frac{S}{2\sqrt{\pi D(t - t')}} e^{-\frac{(x - ut')^2}{4D(t - t')}} \, dt'$$

or, upon changing the variable of integration to $\sigma = \sqrt{t - t'}$ and defining $\zeta := x - ut$,

(4.5-2) $$c = \frac{S}{\sqrt{\pi D}} \int_{0}^{\infty} e^{-\frac{1}{4D}\left(\frac{\zeta}{\sigma} + u\sigma\right)^2} d\sigma.$$

The sign of $\zeta = x - ut$ matters: Positions x with $\zeta > 0$ or $x > ut$ have *not* been crossed by the source before time t, whereas positions with $\zeta < 0$ or $x < ut$ were crossed by the source before time t. This motivates some tricky algebra: We rewrite (4.5-2) as

(4.5-3) $$c = \frac{S}{\sqrt{\pi D}} e^{-\frac{u}{2D}(\zeta + |\zeta|)} \int_{0}^{\infty} e^{-\frac{1}{4D}\left(u\sigma - \frac{|\zeta|}{\sigma}\right)^2} d\sigma.$$

The next change of variables is $\sigma = \sqrt{\frac{|\zeta|}{u}} e^v$, giving

$$c = \frac{S}{\sqrt{\pi D}} e^{-\frac{u}{2D}(\zeta + |\zeta|)} \sqrt{\frac{|\zeta|}{u}} \int_{-\infty}^{\infty} e^{-\frac{u|\zeta|}{D} \sinh^2 v} e^v \, dv$$

$$= \frac{S}{\sqrt{\pi D}} e^{-\frac{u}{2D}(\zeta + |\zeta|)} \sqrt{\frac{|\zeta|}{u}} \int_{-\infty}^{\infty} e^{-\frac{u|\zeta|}{D} \sinh^2 v} \cosh v \, dv.$$

The last equality uses symmetry. The final change of variables is $w = \sinh v$, giving

$$c = \frac{S}{\sqrt{\pi D}} e^{-\frac{u}{2D}(\zeta + |\zeta|)} \sqrt{\frac{|\zeta|}{u}} \int_{-\infty}^{\infty} e^{-\frac{u|\zeta|}{D} w^2} dw,$$

so

$$c = \frac{S}{u} e^{-\frac{u}{2D}(\zeta + |\zeta|)} = \left[\begin{array}{ll} \frac{S}{u} e^{-\frac{u}{D}\zeta}, & \zeta > 0, \\[2mm] \frac{S}{u}, & \zeta < 0. \end{array} \right.$$

The evaluation of the integral representation (4.5-1) is the hard way to do this problem. It is much simpler to solve the ODE boundary value problem for a "traveling wave" solution $c = c(\zeta := x - ut)$. From the diffusion PDE, we derive the ODE

$$uc'(\zeta) + Dc''(\zeta) = 0$$

for $\zeta \neq 0$. The solutions which vanish as $\zeta \to +\infty$, are continuous at $\zeta = 0$ and are bounded as $\zeta \to -\infty$ are

(4.5-4) $$c = \left[\begin{array}{ll} Ce^{-\frac{u}{D}\zeta}, & \zeta > 0, \\[2mm] C, & \zeta < 0. \end{array} \right.$$

Here, C is an arbitrary constant. The source strength S at $\zeta = 0$ is represented by the jump condition on $c'(\zeta)$ at $\zeta = 0$,

$$[Dc'] = -S.$$

Substituting (4.5-4) for c into this jump condition, we determine the constant C as $C = \frac{S}{u}$, and (4.5-4) agrees with the direct evaluation of the integral representation.

The hottest time of the day. Consider the special case of a source localized at the origin for all t, so that $a(t) \equiv 0$. The integral representation (4.20) reduces to

(4.24) $$c(x, t) = \int_{-\infty}^{t} S(t') \frac{e^{-\frac{x^2}{4D(t-t')}}}{2\sqrt{\pi D(t - t')}} \, dt'.$$

Here is a famous problem (Landau & Lifshitz [**17**]) associated with diffusion from a time-dependent source at the origin.

Think of $c(x, t)$ in (4.24) as the temperature field due to a heat source at the origin $x = 0$ with strength proportional to $S(t)$. We take time-periodic $S(t) = 2\sigma \cos \omega t$, and (4.24) becomes

(4.25) $$c(x, t) = \int_{-\infty}^{t} \sigma \cos \omega t' \frac{e^{-\frac{x^2}{4D(t-t')}}}{\sqrt{\pi D(t - t')}} \, dt'.$$

At $x = 0$, this reduces to

$$c(0,t) = \int_{-\infty}^{t} \frac{\sigma \cos \omega t'}{\sqrt{\pi D(t - t')}} \, dt' = \frac{\sigma}{\sqrt{D}} \int_{0}^{\infty} \frac{\cos \omega(t - \tau)}{\sqrt{\pi \tau}} \, d\tau,$$

or

(4.26) $$c(0,t) = \frac{\sigma}{\sqrt{D}} \operatorname{Re} \left\{ e^{i\omega t} \int_{0}^{\infty} \frac{e^{-i\omega \tau}}{\sqrt{\pi \tau}} \, d\tau \right\}.$$

Introduce the change of variables $s = 2\sqrt{\tau}$ in the integral; then it becomes

$$\frac{1}{\sqrt{\pi}} \int_{0}^{\infty} e^{-\frac{i\omega s^2}{4}} \, ds = \sqrt{\frac{1}{i\omega}} = \frac{1}{\sqrt{\omega}} e^{-i\frac{\pi}{4}}$$

for $\omega > 0$. In this case, (4.26) reduces to

(4.27) $$c(0,t) = \frac{\sigma}{\sqrt{D\omega}} \cos \left(\omega t - \frac{\pi}{4} \right).$$

Notice that $c(0,t)$ reaches its maximum one eighth of a cycle later than the source $S(t) = \sigma \cos \omega t$.

Now think of the earth being periodically heated and cooled by the cycle of day and night. Think of $c(x,t)$ for $x > 0$ as the temperature field below the surface. In the solution (4.25), the source feeds diffusion into both half spaces $x > 0$ and $x < 0$, so the restriction of $c(x,t)$ to $x > 0$ represents diffusion along the half line $x > 0$ with source

(4.28) $$-Dc_x(0,t) = \frac{S(t)}{2}$$

at the origin. In the above example we took $S(t) = 2\sigma \cos \omega t$, so (4.28) becomes

(4.29) $$-Dc_x(0,t) = \sigma \cos \omega t.$$

This boundary condition represents the periodic heating and cooling of the earth's surface. The temperature $c(0,t)$ as given by (4.27) says that the hottest time of the day comes one eighth of a cycle, or 3 hours, after high noon.

What about the temperature $c(x,t)$ at depth $x > 0$? In principle, one can evaluate the integral representation (4.25), but it is much harder than the $x = 0$ case. It is more revealing to find a direct solution of the diffusion boundary value problem by separation of variables. Specifically, we seek a time-periodic solution of the diffusion equation $c_t - Dc_{xx} = 0$ in $x > 0$ which vanishes as $x \to \infty$ and satisfies the boundary condition (4.29) at $x = 0$. It is simplest to seek a *complex* solution $c(x,t)$ which satisfies the boundary condition (4.29) with $\cos \omega t$ replaced by $e^{i\omega t}$, and take the real part when we are done. The time-periodic solution of the complex boundary value problem has separation-of-variables form

$$c(x,t) = e^{i\omega t} C(x).$$

Substitution into the diffusion PDE and boundary conditions gives the ODE boundary value problem

$$i\omega C = DC_{xx} \quad \text{in } x > 0,$$
$$C(\infty) = 0,$$
$$DC_x(0) = -\sigma.$$

The ODE has exponential solutions $C = e^{\lambda x}$, where λ satisfies the characteristic equation

$$\lambda^2 = i\frac{\omega}{D}.$$

The roots are

$$\lambda = \pm\sqrt{\frac{\omega}{D}}\, e^{i\pi/4} = \pm\sqrt{\frac{\omega}{D}}\left(\frac{1}{\sqrt{2}} + \frac{i}{\sqrt{2}}\right).$$

The boundary condition $C(\infty) = 0$ selects the root with negative real part, so

$$C = Ae^{-i\left(\sqrt{\frac{\omega}{D}}e^{i\pi/4}\right)x} = Ae^{-\sqrt{\frac{\omega}{D}}\left(\frac{1}{\sqrt{2}} + \frac{i}{\sqrt{2}}\right)x},$$

where A is a constant. A is determined from the boundary condition $DC_x(0) = -\sigma$, so

$$-D\left(-\sqrt{\frac{\omega}{D}}\, e^{i\pi/4}\right) A = -\sigma,$$

or

$$A = \frac{1}{\sqrt{D\omega}}\, e^{-i\pi/4}.$$

Hence,

$$C(x) = \frac{\sigma}{\sqrt{D\omega}}\, e^{-\sqrt{\frac{\omega}{D}}\left(\frac{1}{\sqrt{2}} + \frac{i}{\sqrt{2}}\right)x - i\frac{\pi}{4}},$$

and then

$$(4.30) \qquad c(x,t) = \text{Re}\left\{e^{i\omega t} \frac{\sigma}{\sqrt{d\omega}}\, e^{-\sqrt{\frac{\omega}{D}}\left(\frac{1}{\sqrt{2}} + \frac{i}{\sqrt{2}}\right)x - i\frac{\pi}{4}}\right\}$$

$$= \frac{\sigma}{\sqrt{D\omega}}\, e^{-\sqrt{\frac{\omega}{2D}}x} \cos\left(\omega t - \sqrt{\frac{\omega}{2D}}\,x - \frac{\pi}{4}\right).$$

We see that temperature oscillations decay exponentially with depth, and the e-folding distance is

$$(4.31) \qquad\qquad\qquad\qquad \ell = \sqrt{\frac{2D}{\omega}}.$$

The temperature oscillation at depth x has a phase lag behind the heating,

$$\text{phase lag} = \sqrt{\frac{\omega}{2D}}\, x + \frac{\pi}{4}.$$

At the surface $x = 0$ we get 3 hours; down deep, longer. In fact, the solution (4.31) looks like an exponentially decaying traveling wave, as depicted in Figure 4.1.

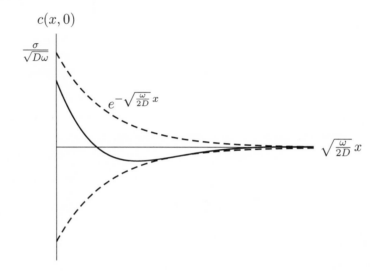

Figure 4.1.

Here is a practical example related to the exponential decay: Indians in North Carolina dug summer lodges deep in the North Carolina clay. They needed to dig deeper than the penetration depth (4.31). How deep is that for $D = 10^{-3}\,\mathrm{cm^2\,s^{-1}}$ (the typical diffusion constant for "dirt") and $\omega = \frac{2\pi}{(360)(24)(60)(60)}\,\mathrm{s^{-1}} = 2 \cdot 10^{-7}\,\mathrm{s^{-1}}$ (yearly cycle)? We get e-folding distance $\ell = 100\,\mathrm{cm} = 1\,\mathrm{m}$.

Problem 4.6 ("Diffusive" damping).   In Figure 4.2, $y = u(x, t)$ is the configuration of a string in the (x, y) plane. The string is strongly overdamped, as if it were immersed in thick syrup, so that $u(x, t)$ satisfies the diffusion PDE with diffusion coefficient D, instead of the usual wave equation. The string is connected to the y-axis. We have $a(t) := u(0, t)$, a given function of time. The string exerts a vertical force $f(t) = T[u_x](0, t)$ at $(x, y) = (0, a(t))$, where T is the string tension. Assuming $u(x, t) \to 0$ as

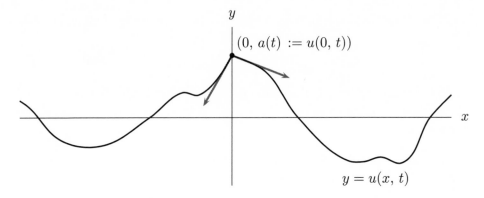

Figure 4.2.

$|x| \to \infty$, we want to determine $f(t)$ as a functional of $a(t)$. We start with the inverse problem, which is more straightforward.

a) Show that

$$a(t) = -\frac{\sqrt{D}}{2T}\frac{1}{\sqrt{\pi}}\int_0^\infty f(t - \frac{s^2}{4})\,ds$$

(assuming that the integral exists).

b) Show that

(4.6-1) $$f(t) = -\frac{2T}{\sqrt{D}}\frac{1}{\sqrt{\pi}}\int_0^\infty \dot{a}(t - \frac{s^2}{4})\,ds.$$

c) The usual linear friction force is proportional to $-\dot{a}(t)$. The "diffusive friction" (4.6-1) is similar, except that it samples $\dot{a}$ at times in the past before the present time t. Now suppose $\dot{a}(t) = v\cos\omega t$, with v and ω being positive constants. The linear friction force is proportional to $-v\cos\omega t = v\cos(\omega t - \pi)$, so its phase lag relative to the velocity is π. What is the phase lag of the "diffusive friction" relative to the velocity?

Solution.

a) A jump in u_x at $x = 0$ implies a point source at $x = 0$ of strength $-D[u_x](0, t) = -\frac{D}{T}f(t)$. The relevant solution to the diffusion PDE with this source strength at the origin is

$$u(x, t) = -\frac{D}{T}\int_{-\infty}^t f(t')\frac{e^{-\frac{x^2}{4D(t-t')}}}{2\sqrt{\pi D(t - t')}}\,dt',$$

and it follows that

$$a(t) = u(0, t) = -\frac{\sqrt{D}}{T} \int_{-\infty}^{t} \frac{f(t')}{2\sqrt{\pi D(t - t')}} \, dt'.$$

Introducing $s := 2\sqrt{t - t'}$ as the variable of integration, this becomes

$$(4.6\text{-}2) \qquad a(t) = -\frac{\sqrt{D}}{2T} \frac{1}{\sqrt{\pi}} \int_{0}^{\infty} f(t - \frac{s^2}{4}) \, ds.$$

b) By use of (4.6-2),

$$(4.6\text{-}3) \qquad \frac{2T}{\sqrt{D}} \frac{1}{\sqrt{\pi}} \int_{0}^{\infty} \dot{a}(t - \frac{s'^2}{4}) \, ds' = \frac{1}{\pi} \int_{0}^{\infty} \int_{0}^{\infty} \dot{f}(t - \frac{s^2}{4} - \frac{s'^2}{4}) \, ds \, ds'.$$

Using (r, ϑ) polar coordinates for the (s, s') plane, the double integral on the right-hand side is

$$\frac{1}{\pi} \int_{0}^{\frac{\pi}{2}} \int_{0}^{\infty} \dot{f}(t - \frac{r^2}{4}) r \, dr \, d\vartheta = \int_{0}^{\infty} \dot{f}(t - \frac{r^2}{4}) \frac{r}{2} \, dr = f(t).$$

Hence, (4.6-3) becomes

$$f(t) = -\frac{2T}{\sqrt{D}} \frac{1}{\sqrt{\pi}} \int_{0}^{\infty} \dot{a}(t - \frac{s'^2}{4}) \, ds'.$$

c) Set $\dot{a}(t) = v e^{i\omega t}$ and compute complex $f(t)$ from (4.6-1). We get

$$f(t) = -\frac{2T}{\sqrt{D}} v \left(\frac{1}{\sqrt{\pi}} \int_{0}^{\infty} e^{-i\frac{\omega s^2}{4}} \, ds \right) e^{i\omega t}$$

$$= -2T \frac{v}{\sqrt{D\omega}} e^{i(\omega t - \frac{\pi}{4})}$$

$$= 2T \frac{v}{\sqrt{D\omega}} e^{i(\omega t - \frac{5\pi}{4})}.$$

Hence, the phase lag of "diffusive friction" is $\frac{5\pi}{4}$, a bit longer than π, because of sampling $-\dot{a}(t)$ in the *past*.

Problem 4.7 (Work against "diffusive friction"). In Problem 4.6, the agent which imposes $a(t) := u(0, t)$ as a given function does work against "diffusive friction" at the rate $-f(t)\dot{a}(t)$. Suppose $a(t)$ is periodic in time with period $\frac{2\pi}{\omega}$. Show that the total work over a cycle is nonnegative.

Solution. The work over one cycle is

$$W = -\int_{0}^{\frac{2\pi}{\omega}} f(t)\dot{a}(t) \, dt.$$

We substitute (4.6-1) for $f(t)$ to get

$$(4.7\text{-}1) \qquad W = \frac{2T}{\sqrt{D}} \frac{1}{\sqrt{\pi}} \int_0^\infty \left(\int_0^{\frac{2\pi}{\omega}} \dot{a}(t) \dot{a}(t - \frac{s^2}{4}) \, dt \right) ds.$$

In the derivation, we interchanged the order of s and t integrations. Next, we substitute for $\dot{a}(t)$ its Fourier series

$$\dot{a}(t) = \sum_{m=-\infty}^{\infty} v_m e^{im\omega t},$$

where $v_{-m} = \bar{v}_m$ for real-valuedness. We compute

$$\dot{a}(t) \dot{a}(t - \frac{s^2}{4}) = \sum_{m,n} v_m v_n e^{i\omega(m+n)t - in\omega \frac{s^2}{4}},$$

and the t-integral is

$$(4.7\text{-}2) \qquad
\begin{aligned}
\int_0^{\frac{2\pi}{\omega}} \dot{a}(t) \dot{a}(t - \frac{s^2}{4}) \, dt &= \frac{2\pi}{\omega} \sum_m v_m v_{-m} e^{im\omega \frac{s^2}{4}} \\
&= \frac{4\pi}{\omega} \sum_{m \geq 0} |v_m|^2 \cos m\omega \frac{s^2}{4}.
\end{aligned}$$

The last equality uses $v_{-m} = \bar{v}_m$. We now apply the s-integration to (4.7-2):

$$
\begin{aligned}
\frac{1}{\sqrt{\pi}} \int_0^\infty \int_0^{\frac{2\pi}{\omega}} \dot{a}(t) \dot{a}(t - \frac{s^2}{4}) \, dt \, ds &= \frac{4\pi}{\omega} \sum_{m \geq 0} |v_m|^2 \frac{1}{\sqrt{\pi}} \int_0^\infty \cos m\omega \frac{s^2}{4} \, ds \\
&= \frac{4\pi}{\omega} \frac{1}{\sqrt{2\omega}} \sum_{m \geq 0} \frac{|v_m|^2}{\sqrt{m}}.
\end{aligned}$$

Hence, (4.7-1) becomes

$$W = 4\pi \frac{T}{\omega} \sqrt{\frac{2}{D\omega}} \sum_{m \geq 0} \frac{|v_m|^2}{\sqrt{m}},$$

and the right-hand side is clearly nonnegative.

A more physical calculation connects the work to the displacement field $u(x,t)$ of the string. First, we have

$$-f(t)\dot{a}(t) = -T\,[u_x](0,t) u_t(0,t).$$

The time-periodic solution for $u(x,t)$ which vanishes as $|x| \to \infty$ is *even* in x, so $[u_x](0,t) = 2u_x(0^+,t)$ and hence

$$-f(t)\dot{a}(t) = -2T(u_x u_t)(0^+,t).$$

Therefore, the work over a cycle is

$$(4.7\text{-}3) \qquad W = -2T \int_0^{\frac{2\pi}{\omega}} (u_x u_t)(0^+, t)\, dt = 2T \int_0^{\frac{2\pi}{\omega}} \int_0^\infty (u_x u_t)_x\, dx\, dt.$$

In the right-hand side,

$$(u_x u_t)_x = u_{xx} u_t + u_x u_{xt} = \frac{1}{D} u_t^2 + \frac{1}{2}(u_x^2)_t.$$

The last equality uses the diffusion PDE to substitute $\frac{1}{D} u_t$ for u_{xx}. Hence, (4.7-3) becomes

$$W = \frac{2T}{D} \int_0^{\frac{2\pi}{\omega}} \int_0^\infty u_t^2\, dx + T \int_0^\infty \int_0^{\frac{2\pi}{\omega}} (u_x^2)_t\, dt\, dx.$$

The second term on the right-hand side vanishes by the time periodicity of u_x^2. This leaves

$$W = \frac{2T}{D} \int_0^{\frac{2\pi}{\omega}} \int_0^\infty u_t^2\, dx\, dt \geq 0.$$

Problem 4.8 (Oscillatory point source solution in $\mathbb{R}^3$ and descent to $\mathbb{R}^2$).

a) Construct the solution to the diffusion PDE in $\mathbb{R}^3$ minus the origin induced by a time-periodic point source at the origin with strength $S \cos \omega t$.

b) Construct the analogous point source solution in $\mathbb{R}^2$ minus the origin by superposition of $\mathbb{R}^3$ point source solutions along the z-axis.

c) Alternatively, we attempt to construct the $\mathbb{R}^2$ point source solution by seeking a separation-of-variables solution $c = \gamma(r)e^{i\omega t}$ of the $\mathbb{R}^2$ equations. Formulate the boundary value problem for $\gamma(r)$ and propose an integral representation of its solution based on comparison with your result in Part b.

Solution.

a) We compute a complex solution due to the source strength $Se^{i\omega t}$, with the understanding that we will take the real part when we are done. Substituting the spherically symmetric separation-of-variables solution

$$(4.8\text{-}1) \qquad c = C\left(\rho := \sqrt{x^2 + y^2 + z^2} \right) e^{i\omega t}$$

into the spherically symmetric $\mathbb{R}^3$ diffusion PDE

$$(\rho c)_t - D(\rho c)_{\rho\rho} = 0$$

gives the ODE

(4.8-2) $$D(\rho C)'' - i\omega(\rho C) = 0,$$

and the solutions which decay as $\rho \to \infty$ are

(4.8-3) $$C = \frac{C_0}{\rho} e^{-\sqrt{\frac{\omega}{2D}}(1+i)\rho},$$

where C_0 is an arbitrary constant. Substituting (4.8-1) into the flux condition

(4.8-4) $$-D \lim_{\varepsilon \to 0} 4\pi\varepsilon^2 c_\rho(\varepsilon, t) = S e^{i\omega t}$$

gives

(4.8-5) $$-D \lim_{\varepsilon \to 0} 4\pi\varepsilon^2 C'(\varepsilon) = S.$$

Next, substituting (4.8-3) into (4.8-5) determines the constant C_0 as

(4.8-6) $$C_0 = \frac{S}{4\pi D}.$$

It now follows from (4.8-1), (4.8-3) and (4.8-6) that

$$c(\rho, t) = \frac{S}{4\pi D\rho} e^{i\omega t - \sqrt{\frac{\omega}{2D}}(1+i)\rho},$$

and the real-valued solution is

(4.8-7) $$c(\rho, t) = \frac{S}{4\pi D\rho} e^{-\sqrt{\frac{\omega}{2D}}\rho} \cos(\omega t - \sqrt{\frac{\omega}{2D}}\rho).$$

b) The line source along the z-axis has strength per unit length $S e^{i\omega t}$. The solution in $\mathbb{R}^3$ induced by this line source has cylindrical symmetry about the z-axis, that is, $c = c(r := \sqrt{x^2 + y^2}, t)$. The flux condition on $c(r, t)$ due to the line source is

(4.8-8) $$-D \lim_{\varepsilon \to 0} 2\pi\varepsilon c_r(\varepsilon, t) = S e^{i\omega t}$$

in place of (4.8-4). Evidently, $c(r, t)$ is the required $\mathbb{R}^2$ point source solution. We construct $c(r, t)$ by *superposition* of $\mathbb{R}^3$ point source solutions with their sources along the z-axis. It is sufficient to evaluate the superposition in the (x, y) plane:

(4.8-9) $$c(r, t) = \frac{S}{4\pi D} \int_{-\infty}^{\infty} \frac{e^{-\sqrt{\frac{\omega}{2D}}(1+i)\sqrt{r^2+z^2}+i\omega t}}{\sqrt{r^2 + z^2}} \, dz.$$

The integrand is the $\mathbb{R}^3$ point source solution due to the source at $z e_z$, evaluated at distance r from the origin of the (x, y) plane. The real-valued $\mathbb{R}^2$ solution is

$$c(r, t) = \frac{S}{4\pi D} \int_{-\infty}^{\infty} \frac{e^{-\sqrt{\frac{\omega}{2D}}\sqrt{r^2+z^2}}}{\sqrt{r^2 + z^2}} \cos(\omega t - \sqrt{\frac{\omega}{2D}}\sqrt{r^2 + z^2}) \, dz.$$

c) The ODE for $\gamma(r)$ is

(4.8-10)
$$D(\gamma'' - \frac{1}{r}\gamma') - i\omega\gamma = 0$$

in $r > 0$. We seek the solution which vanishes as $r \to \infty$ and satisfies

(4.8-11)
$$-D \lim_{\varepsilon \to 0} 2\pi\varepsilon\gamma'(\varepsilon) = S,$$

which follows from substitution of $c = \gamma(r)e^{i\omega t}$ into the $\mathbb{R}^2$ flux condition (4.8-8). If we were in $\mathbb{R}^3$, the coefficient of γ' in the ODE (4.8-10) would be $2\frac{D}{r}$ instead of $\frac{D}{r}$, and we would be right back to (4.8-2) (with replacements of γ for C and r for ρ). The solutions of (4.8-10)–(4.8-11) are *not* expressible in terms of elementary functions. They are in fact a subspecies of Bessel functions. Here, we determine an integral representation of $\gamma(r)$ by comparing $\gamma(r)e^{i\omega t}$ with the superposition (4.8-9). Removing the time factor $e^{i\omega t}$ from the latter, we recognize that

(4.8-12)
$$\gamma(r) = \frac{S}{4\pi D} \int_{-\infty}^{\infty} \frac{e^{-\sqrt{\frac{\omega}{2D}}(1+i)\sqrt{r^2+z^2}}}{\sqrt{r^2 + z^2}} \, dz.$$

The formula (4.8-12) is in fact an integral representation of one of these aforementioned Bessel functions.

Problem 4.9 (Direct analysis of the $\mathbb{R}^2$ integral representation).

a) From the integral representation of $\gamma(r)$ in Problem 4.8c, find a simple approximation to $\gamma(r)$ for $r \gg \sqrt{\frac{D}{\omega}}$. What is the phase lag of $c(r,t)$ relative to the oscillating source at $r = 0$ for $r \gg \sqrt{\frac{D}{\omega}}$?

b) Show that $\gamma(r)$ in Problem 4.8c satisfies the $\mathbb{R}^2$ flux condition

$$\lim_{\varepsilon \to 0} (2\pi\varepsilon)(-D\gamma'(\varepsilon)) = S.$$

Solution.

a) The factor $\sqrt{r^2 + z^2}$ suggests the hyperbolic trigonometric substitution $z = r \sinh \zeta$, and (4.8-12) becomes the ζ-integral

(4.9-1)
$$\gamma(r) = \frac{S}{4\pi D} \int_{-\infty}^{\infty} e^{-zr \cosh \zeta} \, d\zeta,$$

where

(4.9-2)
$$z := \sqrt{\frac{\omega}{2D}}(1+i).$$

Expression (4.9-2) is a good representation for analyzing the $r \gg \sqrt{\frac{D}{\omega}}$ limit.
For $r \gg \sqrt{\frac{D}{\omega}}$, $|z|r \gg 1$ and the main contribution to the integral (4.9-1)
comes from a small neighborhood of $\zeta = 0$, where the two-term Taylor
expansion $\cosh \zeta \simeq 1 + \frac{\zeta^2}{2}$ applies. Hence, (4.9-1) is replaced by a simple
Gaussian integral, and its evaluation gives

(4.9-3)
$$\gamma(r) \simeq \frac{1}{2}\frac{S}{D}\frac{e^{-zr}}{\sqrt{2\pi z r}}.$$

Now insert the explicit value (4.9-2) of z. We have

$$\sqrt{zr} = \sqrt{\sqrt{\frac{\omega r^2}{D}}\frac{1+i}{\sqrt{2}}} = \left(\frac{\omega r^2}{D}\right)^{\frac{1}{4}}e^{i\frac{\pi}{8}},$$

and (4.9-3) becomes

$$\gamma(r) \simeq \frac{1}{2\sqrt{2\pi}}\frac{S}{D}\left(\frac{D}{\omega r^2}\right)^{\frac{1}{4}}e^{-\sqrt{\frac{\omega}{2D}}r}e^{-i(\sqrt{\frac{\omega}{2D}}r+\frac{\pi}{8})}.$$

Next, multiply by $e^{i\omega t}$ and take the real part to recover the $\mathbb{R}^2$ "oscillatory
source" solution of the diffusion PDE,

$$c(r,t) \simeq \frac{1}{2\sqrt{2\pi}}\frac{S}{D}\left(\frac{D}{\omega r^2}\right)^{\frac{1}{4}}e^{-\sqrt{\frac{\omega}{2D}}r}\cos\left(\omega t - \sqrt{\frac{\omega}{2D}}r - \frac{\pi}{8}\right),$$

and we see that

$$\text{phase lag} \simeq \sqrt{\frac{\omega}{2D}}r + \frac{\pi}{8}$$

for $r \gg \sqrt{\frac{D}{\omega}}$. Let us put this in context: For the $\mathbb{R}^3$ oscillatory source
solution (4.8-7), the phase lag is $\sqrt{\frac{\omega}{2D}}r$. In $\mathbb{R}^1$, we found the phase lag to be
$\sqrt{\frac{\omega}{2D}}x + \frac{\pi}{4}$ at distance x from the source. Our $\mathbb{R}^2$ result with the $\frac{\pi}{8}$ phase
shift seems to be in the middle.

b) The ζ-integral representation (4.9-1), which served so well for $r \gg \sqrt{\frac{D}{\omega}}$,
needs to be re-examined in the limit $r \to 0$. We use even symmetry in ζ to
write $\gamma(r)$ as an integral over $\zeta > 0$, and then perform a second change of
variable, $v = re^\zeta$. Thus, (4.9-1) becomes the v-integral

$$\gamma(r) = \frac{S}{2\pi D}\int_r^\infty \frac{1}{v}e^{-z(\frac{v}{2}+\frac{r^2}{2v})}\,dv.$$

Differentiating with respect to r, we get

$$\gamma'(r) = -\frac{S}{2\pi Dr}e^{-zr} - \frac{Sr}{2\pi D}\int_r^\infty \frac{e^{-z(\frac{v}{2} + \frac{r^2}{2v})}}{v^2}\, dv.$$

The second term on the right-hand side is bounded in absolute value by

$$\frac{Sr}{2\pi D}\int_r^\infty \frac{dr}{r^2} = \frac{S}{2\pi D},$$

so

$$(2\pi\varepsilon)(-D\gamma'(\varepsilon)) = Se^{-z\varepsilon} + O(\varepsilon) = S + O(\varepsilon)$$

as $\varepsilon \to 0$. Here, $O(\varepsilon)$ means a term bounded in absolute value by a positive constant times ε. The $\mathbb{R}^2$ flux condition is immediate.

Guide to bibliography. Recommended references for this chapter are Courant & Hilbert [**8, 9**], Evans [**11**], Feynman volume II [**13**], John [**15**], and Landau & Lifshitz [**17**].

The prime reference is Evans [**11**] because the standard constructions of solutions based on linearity and superposition are concentrated together and clearly presented in Chapter 2 of his book. In particular, one finds Green's identity constructions analogous to our Problem 4.2, and the method of descent is applied to the wave equation. Courant & Hilbert [**8, 9**] and John [**15**] contain this material but in a more dispersed form. Courant & Hilbert present useful standard calculations, such as reducing spherical waves and diffusions to one-dimensional waves and diffusions, or eliminating first space derivatives from a steady convection-diffusion PDE (used in analysis of the smoke plume, Problem 4.3). Feynman in volume II [**13**] evokes the superposition principle to simply write down the superposition of point charge solutions that solves Poisson's equation. This, the standard practice of physicists, is how we began our chapter. Finally, Landau & Lifshitz [**17**] present time-periodic, spatially attenuated wave solutions of the diffusion PDE.

δ-functions

In the prototype examples of Chapter 4, there is the implicit notion of a source $s(\boldsymbol{x})$ whose region of support in $\mathbb{R}^n$ is smaller than we care to resolve, but whose total strength $\int_{\mathbb{R}^n} s(x)\, dx$ remains nonzero. In the analysis of those problems, we did not attempt to describe a "point" source as an actual function of spacetime variables. But there is a tradition in physics of doing just that, embodied in the formal notion of the *Dirac δ-function*: Let $\boldsymbol{x}$ in $\mathbb{R}^n$ represent position in space or events in spacetime. The "definition" of the Dirac δ-function is that

(5.1) $$\delta(\boldsymbol{x}) = 0 \quad \text{for } \boldsymbol{x} \neq 0$$

but nevertheless

(5.2) $$\int_{\mathbb{R}^n} \delta(\boldsymbol{x})\, d\boldsymbol{x} = 1$$

for any region R that contains the origin $\boldsymbol{x} = 0$. As mathematicians know, there is no such function. We examine two kinds of answer as to what is actually meant.

First, δ-functions are part of a shorthand: formal notation and procedures for the nonrigorous but intuitively clear solution of ODE and PDE boundary value problems. As the problems become harder and deeper, the effectiveness of the δ-function "shorthand" becomes clear. The advantages of brevity and insight are substantial.

Second, one recognizes δ-*sequences* arising from problems in analysis, and analysis of linear PDEs in particular. There exist ε-sequences of functions $f(\boldsymbol{x}, \varepsilon)$, $\boldsymbol{x}$ in $\mathbb{R}^n$ and $\varepsilon > 0$, such that for any bounded continuous function $f(x)$,

(5.3) $$\lim_{\varepsilon \to 0} \int_{\mathbb{R}^n} f(x)\delta(x, \varepsilon)\, dx = f(0).$$

One is tempted to say, "that is easy, just let the support of $\delta(x, \varepsilon)$ shrink to the single point $x = 0$ as $\varepsilon \to 0$, while maintaining $\int_{\mathbb{R}^n} \delta(x, \varepsilon)\, dx = 1$." But it is *not* that easy: There are legions of $\delta(x, \varepsilon)$ arising from "ordinary" problems which satisfy (5.3) but have no pointwise limit as $\varepsilon \to 0$ with $x \neq 0$ fixed.

δ-functions as "shorthand". The ground rule is: "When you see $\delta(x)$, integrate over a region R containing $x = 0$, use $\int_R \delta(x)\, dx = 1$, and pretend that the rules of calculus work." Retaining your common sense, you can navigate this "δ-function shorthand" without a long list of formalized rules.

Let us start with the one-dimensional sheet source problem (4.2)–(4.5). The shorthand presentation of it is the ODE

(5.4) $$c'' - c = -\delta(x)$$

for "all" x (including $x = 0$) together with the zero boundary condition as $|x| \to \infty$. Assuming c is bounded, integrate (5.4) over $-\varepsilon < x < \varepsilon$, use $\int_{-\varepsilon}^{\varepsilon} \delta(x)\, dx = 1$, and take the limit $\varepsilon \to 0$. We find the jump condition on c' at $x = 0$, $[c'] = -1$. Given c' with a simple jump discontinuity at $x = 0$, we expect

$$c(x) = c_0 + \int_0^x c'(s)\, ds$$

for some constant c_0. Hence, $c(0^-) = c(0^+) = c_0$, a common value; so c is continuous at $x = 0$, i.e., $[c] = 0$. In summary, we have replaced the δ-function on the right-hand side of (5.4) with the two jump conditions $[c'] = -1$ and $[c] = 0$, and we are back to the original formulation (4.2)–(4.5).

For the "point source" problem in $\mathbb{R}^3$, $c(x)$ satisfies the PDE

(5.5) $$\Delta c - c = -\delta(x),$$

again for "all" x, including $x = 0$. We integrate (5.5) over a ball of radius ε about the origin and use the divergence theorem. We get

$$\int_{r=\varepsilon} c_r\, da - \int_{r<\varepsilon} c\, dx = -1.$$

Now recall that the solution for c blows up like $\frac{1}{r}$ as $r \to 0$, so the integral $\int_{r<\varepsilon} c\, dx$ has an upper bound equal to a constant times $\int_0^\varepsilon (\frac{1}{r}) r^2\, dr = \frac{\varepsilon^2}{2}$.

Hence,

(5.6) $$\int_{r<\varepsilon} c\,d\boldsymbol{x} \to 0 \quad \text{as } \varepsilon \to 0.$$

In our δ-function shorthand we don't really have an a priori argument for (5.6). The policy is: "Assume it. Compute your solution, and do an a posteriori consistency check of (5.6)." In summary, we have replaced the δ-function in (5.5) by the flux condition

$$\lim_{\varepsilon \to 0} \int_{r=\varepsilon} c_r\,da = -1,$$

and the analysis proceeds from here just as in Chapter 4.

The δ-function shorthand "explanation" of superposition is a formal calculation. Roughly, it says: "Any source is a sum of point sources, and the field generated by the source is a corresponding (same weights) sum of point source fields." For instance, let us consider

$$\Delta c - c = -s(\boldsymbol{x})$$

as in (4.11). We divide $\mathbb{R}^3$ into little volume elements $d\boldsymbol{x}'$. The source function which equals $s(\boldsymbol{x}')$ in the volume element $d\boldsymbol{x}'$ about $\boldsymbol{x}'$ and is zero otherwise looks like a point source $(s(\boldsymbol{x}')\,d\boldsymbol{x}')\delta(\boldsymbol{x} - \boldsymbol{x}')$. The sum of these point sources over all volume elements $d\boldsymbol{x}'$ should formally recover $s(\boldsymbol{x})$, so we have the "δ-function identity"

(5.7) $$s(\boldsymbol{x}) = \int_{\mathbb{R}^n} s(\boldsymbol{x}')\delta(\boldsymbol{x} - \boldsymbol{x}')\,d\boldsymbol{x}'.$$

Alternatively, you might say: "The support of $\delta(\boldsymbol{x} - \boldsymbol{x}')$ in $\boldsymbol{x}'$ is a tiny neighborhood of $\boldsymbol{x}' = \boldsymbol{x}$, and there $s(\boldsymbol{x}')$ differs negligibly from the uniform value $s(\boldsymbol{x})$. We pull $s(\boldsymbol{x})$ outside the integral and (5.7) is obvious by (5.2)." Now here is the "explanation" of the superposition solution

(5.8) $$c(\boldsymbol{x}) := \int_{\mathbb{R}^3} s(\boldsymbol{x}')g(\boldsymbol{x} - \boldsymbol{x}')\,d\boldsymbol{x}',$$

where

(5.9) $$g(\boldsymbol{x}) := \frac{e^{-|\boldsymbol{x}|}}{4\pi|\boldsymbol{x}|}.$$

The $g(\boldsymbol{x})$ in (5.9) is the point source solution which satisfies

$$\Delta g - g = -\delta(\boldsymbol{x}).$$

For $c(\boldsymbol{x})$ in (5.8), we formally calculate

$$(\Delta c - c)(\boldsymbol{x}) = \int_{\mathbb{R}^3} s(\boldsymbol{x}')(\Delta g - g)(\boldsymbol{x} - \boldsymbol{x}')\,d\boldsymbol{x}' = -\int_{\mathbb{R}^3} s(\boldsymbol{x}')\delta(\boldsymbol{x} - \boldsymbol{x}')\,d\boldsymbol{x}',$$

and the right-hand side is $-s(\boldsymbol{x})$ by (5.7). A similar "explanation" of the superposition solution (4.19) of the inhomogeneous diffusion PDE (4.18) is based on the diffusion kernel $g(x,t)$ in (2.39), which satisfies

$$g_t - Dg_{xx} = \delta(x,t).$$

Problem 5.1 (Correctly representing a point load).   An elastic cord in the (x,y) plane is suspended between $(0,0)$ and $(L,0)$. Its configuration is represented by the graph of $y = y(x)$, with $y(0) = y(L) = 0$. The cord is under uniform tension T (a force) and subject to a vertical load $-f(x)$ per unit length.

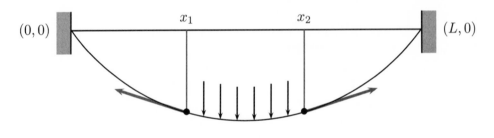

Figure 5.1.

a)   Derive an ODE boundary value problem satisfied by $y(x)$. We want a fully nonlinear problem. Think of Figure 5.1: The little vertical arrows represent the vertical load $-f(x)$ between $x = x_1$ and $x = x_2$, and the tangential "tension" arrows at $x = x_1$ and $x = x_2$ have magnitude T.

b)   Determine $y(x)$ for a "point" load F at $x = x'$, where $0 < x' < L$. In particular, determine an equation for $y_* := y(x')$ and solve it asymptotically in the small-deflection limit $\frac{F}{T} \to 0$.

Solution.

a)   The total vertical load between $x = x_1$ and $x = x_2$ is

(5.1-1) $$-\int_{x_1}^{x_2} f(x)\sqrt{1+y_x^2}\, dx.$$

This is balanced by the vertical components of tension at $x = x_1$ and $x = x_2$. The sum of the vertical components is

$$(5.1\text{-}2) \qquad T \left[\frac{y_x}{\sqrt{1+y_x^2}} \right]_{x_1}^{x_2} = T \int_{x_1}^{x_2} \left(\frac{y_x}{1+y_x^2} \right)_x dx.$$

Equating (5.1-1) and (5.1-2), we have

$$\int_{x_1}^{x_2} \left\{ T \left(\frac{y_x}{\sqrt{1+y_x^2}} \right)_x + f\sqrt{1+y_x^2} \right\} dx = 0.$$

This holds for all x_1 and x_2 with $0 < x_1 < x_2 < L$, so $y(x)$ satisfies the ODE

$$(5.1\text{-}3) \qquad \frac{1}{\sqrt{1+y_x^2}} \left(\frac{y_x}{\sqrt{1+y_x^2}} \right)_x = -\frac{f}{T}$$

in $0 < x < L$. The boundary conditions are of course $y(0) = y(L) = 0$.

b) In $0 < x < x'$ or $x' < x < L$, $f(x) = 0$ and it follows from (5.1-3) that y_x is uniform. The uniform values in $0 < x < x'$ and $x' < x < L$ are generally different. In view of the zero boundary conditions at $x = 0$, L and the continuity of y at x', we surmise that

$$(5.1\text{-}4) \qquad y(x) = \left[\begin{array}{ll} y_* \frac{x}{x'}, & 0 < x < x', \\[2ex] y_* \frac{L-x}{L-x'}, & x' < x < L. \end{array} \right.$$

The value of $y_* := y(x')$ is determined by an appropriate jump condition on y_x at $x = x'$.

You might think that the point load is represented by

$$(5.1\text{-}5) \qquad f(x) = F\delta(x - x').$$

Substitute this point load into the ODE (5.1-3), multiply by $\sqrt{1+y_x^2}$ and integrate over $x' - \varepsilon < x < x' + \varepsilon$ to get

$$\left[\frac{y_x}{\sqrt{1+y_x^2}} \right]_{x_1}^{x_2} = -\frac{F}{T} \int_{x'-\varepsilon}^{x'+\varepsilon} \sqrt{1+y_x^2} \, \delta(x - x') \, dx.$$

Look at the right-hand side. If $\sqrt{1+y_x^2}$ were continuous at $x = x'$, with some value σ, we would gladly replace the right-hand side by $-\frac{F}{T}\sigma$. But we anticipate that y_x and hence $\sqrt{1+y_x^2}$ has a jump discontinuity *exactly* at $x = x'$, where the δ-function is "supported". Do we take the average of the left-hand and right-hand limits of $\sqrt{1+y_x^2}$? Or maybe some other weighted average? No and no.

It turns out that (5.1-5) *misrepresents* the point load: Let s be the *arc length* of cord between $(0,0)$ and $(x, y(x))$. Let s' be the arc length at x'.

Since f is *load per unit length*, our point load *should* have been represented by

(5.1-6) $$f = F\delta(s - s').$$

Our jump condition on y_x should be determined by integration with respect to s from $s' - \varepsilon$ to $s' + \varepsilon$. First, convert the independent variable in the ODE (5.1-3) from x to s. We have $s_x = \sqrt{1 + y_x^2}$, so

$$\partial_s = \frac{1}{s_x}\partial_x = \frac{1}{\sqrt{1 + y_x^2}}\partial_x,$$

and (5.1-3) with the point load (5.1-6) reads

$$y_{ss} = -\frac{F}{T}\delta(s - s').$$

Hence, we have the jump condition $[y_s](s') = -\frac{F}{T}$ or, converting back to the x variable,

(5.1-7) $$\left[\frac{y_x}{\sqrt{1 + y_x^2}}\right](x') = -\frac{F}{T}.$$

Notice that this is what a "common sense" force balance like in Part a would give. Substituting (5.1-4) into (5.1-7), we obtain an equation for $y_* := y(x')$,

$$y_*\left\{\frac{1}{\sqrt{(L - x')^2 + y_*^2}} + \frac{1}{\sqrt{x'^2 + y_*^2}}\right\} = -\frac{F}{T}.$$

In the small-deflection limit with $\frac{F}{T}$, $y_* \to 0$, this reduces asymptotically to

$$\frac{y_*}{L} = -\frac{F}{T}\frac{x'}{L}\left(1 - \frac{x'}{L}\right).$$

Problem 5.2 (Correlation function of colored noise).   We consider an ensemble of one-dimensional Brownian motions. Each $x(t)$ starts from the origin at time zero, i.e., $x(0) = 0$. As we know, the distribution of $x(t)$ at any time $t > 0$ is given by the diffusion kernel. But now we look at the velocities, $w(t) := \dot{x}(t)$. The ensemble of $w(t)$'s is called *white noise*. The statistics of white noise is

(5.2-1) $$\langle w(t)\rangle = 0,$$

(5.2-2) $$\langle w(t)w(t')\rangle = W\delta(t - t'),$$

where the brackets $\langle\ \rangle$ denote ensemble averaging and W in (5.2-2) is a positive constant. The intuitive idea of the *correlation function* $\langle w(t)w(t')\rangle$ in (5.2-2) is this: For $t \neq t'$, there is complete statistical independence of

$w(t)$ and $w(t')$, so $\langle w(t)w(t')\rangle = \langle w(t)\rangle\langle w(t')\rangle = 0$. If each $w(t)$ were an ordinary function of time, generally nonzero, we would have $\langle w^2(t)\rangle > 0$ and so the support of $\langle w(t)w(t')\rangle$ ought to be the line $t = t'$. This suggests proportionality to $\delta(t - t')$ as the natural choice.

a) If the diffusion coefficient of the Brownian motion is D, what is W in (5.2-2)?

b) We now construct an ensemble of $c(t)$, called *colored noise*. Each $c(t)$ in the colored noise ensemble is related to a $w(t)$ in a white noise ensemble by

(5.2-3) $$\varepsilon\dot{c}(t) + c(t) = w(t).$$

This time, the $w(t)$ live in $-\infty < t < \infty$, and the statistics (5.2-1) and (5.2-2) remain in force. ε is a positive constant. Formally, we expect that colored noise asymptotes to white noise as $\varepsilon \to 0$.

Pretending that each $w(t)$ is an ordinary bounded function, show that solutions to (5.2-3) asymptote to

(5.2-4) $$c(t) = \frac{1}{\varepsilon}\int_0^\infty e^{-s}w(t - s)\,ds$$

as $t \to \infty$. Next, we take (5.2-4) as the unique mapping from $w(t)$ to $c(t)$. Compute $\langle c(t)\rangle$ and $\langle c(t)c(t')\rangle$.

Solution.

a) The mapping from white noise to Brownian motion is integration,

$$x(t) = \int_0^t w(s)\,ds.$$

Hence,

$$\langle x^2(t)\rangle = \int_0^t \int_0^t \langle w(s)w(s')\rangle\,ds'ds$$
$$= W\int_0^t \int_0^t \delta(s - s')\,ds'ds$$
$$= W\int_0^t ds$$
$$= Wt$$

and, comparing with $\langle x^2(t)\rangle = 2Dt$, we see that $W = 2D$.

b) The ODE (5.2-3) is equivalent to

$$\varepsilon(e^{\frac{t}{\varepsilon}}c(t))' = e^{\frac{t}{\varepsilon}}w(t).$$

Assuming $w(t)$ to be bounded, there is a particular solution

$$c(t) = \frac{1}{\varepsilon} \int_{-\infty}^{t} e^{-\frac{t-u}{\varepsilon}} w(u) \, du,$$

and changing the variable of integration from u to $s = t - u$ converts this solution to (5.2-4). The general solution of the ODE (5.2-3) is the particular solution of (5.2-3) plus any multiple of $e^{t/\varepsilon}$. This exponential decays as $t \to \infty$, leaving the particular solution.

Now for the statistics of colored noise. $\langle c(t) \rangle = 0$ is obvious by (5.2-1) and the linearity of the $w \to c$ mapping (5.2-4). We compute the c-correlation function as follows:

$$\langle c(t)c(t') \rangle = \frac{1}{\varepsilon^2} \int_0^\infty \int_0^\infty e^{-\frac{s+s'}{\varepsilon}} \langle w(t-s)w(t'-s') \rangle \, ds' ds$$

(5.2-5)

$$= \frac{2D}{\varepsilon^2} \int_0^\infty \int_0^\infty e^{-\frac{s+s'}{\varepsilon}} \delta(t - t' - (s - s')) \, ds' ds.$$

In (5.2-5), the region of integration is the first quadrant of the (s, s') plane, and the support of the δ-function is the line $s' = s + (t' - t)$. In Figure 5.2, we have graphed this line for $t' > t$.

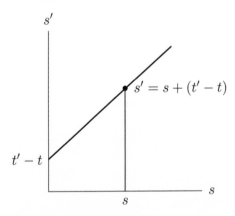

Figure 5.2.

For $t' > t$, we first do the s' integral, then the s integral, obtaining

$$\langle c(t)c(t') \rangle = \frac{2D}{\varepsilon} e^{-\frac{t'-t}{\varepsilon}} \int_0^\infty e^{-\frac{2s}{\varepsilon}} \, ds = \frac{D}{\varepsilon} e^{-\frac{t'-t}{\varepsilon}}.$$

We do $t < t'$ by a similar calculation, with the order of integration reversed. In summary,

(5.2-6) $$\langle c(t)c(t') \rangle = \frac{D}{\varepsilon} e^{-\frac{1}{\varepsilon}|t-t'|}.$$

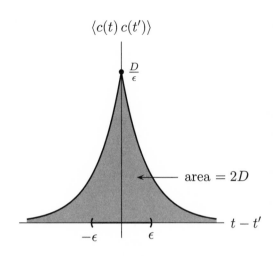

$$\langle c(t)\, c(t')\rangle$$

$$\frac{D}{\epsilon}$$

area $= 2D$

$$t - t'$$

$-\epsilon \qquad \epsilon$

Figure 5.3.

Recall that colored noise is supposed to be "asymptotic to white noise" as $\varepsilon \to 0$. In what sense does (5.2-6) "reduce" to $2D\delta(t - t')$ as $\varepsilon \to 0$? In Figure 5.3, we have plotted $\langle c(t)c(t')\rangle$ versus $t - t'$ for a "small" ε. The support of $\langle c(t)c(t')\rangle$ is concentrated in an interval of thickness ε about $t - t' = 0$, and the integral over all $t - t'$ is $2D$. Thus, (5.2-6) is a simple example of a δ-*sequence* for $2D\delta(t - t')$.

It is now time to reveal the full power of the δ-function "shorthand": Suppose we have a point source of heat in $\mathbb{R}^3$ which has constant strength s, but whose position is a function of time,

$$(5.10) \qquad\qquad \boldsymbol{x} = \boldsymbol{a}(t).$$

$\boldsymbol{a}(t)$ is confined to the ball of radius ε about the origin for all time. The formal PDE for the heat content $h(\boldsymbol{x}, t)$ is

$$(5.11) \qquad h_t - D\Delta h = s \int_{-\infty}^{\infty} \delta(\boldsymbol{x} - \boldsymbol{a}(t'), t - t')\, dt.$$

To examine the h-field far from the origin, $|\boldsymbol{x}| \gg \varepsilon$, we formally Taylor expand (!) the δ-function in $\boldsymbol{x}$, and (5.11) becomes

$$(5.12) \quad h_t - D\Delta h = s \int_{-\infty}^{\infty} \left\{ \delta(\boldsymbol{x}, t - t') - \boldsymbol{a}(t') \cdot \boldsymbol{\nabla}\delta(\boldsymbol{x}, t - t') \right\} dt' + O(\varepsilon^2).$$

Now observe that

$$f(\boldsymbol{x}) := \int_{-\infty}^{\infty} \delta(\boldsymbol{x}, t - t')\, dt'$$

is the *spatial* δ-function on $\mathbb{R}^3$, that is, $f(\boldsymbol{x}) = \delta(\boldsymbol{x})$: For $\boldsymbol{x} \neq 0$, we have $\delta(\boldsymbol{x}, t - t') = 0$ for all t', so $f(\boldsymbol{x}) = 0$ for $\boldsymbol{x} \neq 0$; also,

$$\int_{\mathbb{R}^3} f(\boldsymbol{x})\, d\boldsymbol{x} = \int_{\mathbb{R}^3} \int_{-\infty}^{\infty} \delta(\boldsymbol{x}, t - t')\, dt' = 1.$$

We rewrite (5.12) as

$$(5.13) \qquad h_t - D\Delta h = s\delta(\boldsymbol{x}) - s\boldsymbol{a}(t) \cdot \boldsymbol{\nabla}\delta(\boldsymbol{x}) + O(\varepsilon^2).$$

Here, we have used another formal calculation,

$$\int_{-\infty}^{\infty} \boldsymbol{a}(t') \cdot \boldsymbol{\nabla}\delta(\boldsymbol{x}, t - t') \, dt' = \boldsymbol{a}(t) \cdot \int_{-\infty}^{\infty} \boldsymbol{\nabla}\delta(\boldsymbol{x}, t - t') \, dt'$$

$$= \boldsymbol{a}(t) \cdot \boldsymbol{\nabla} \int_{-\infty}^{\infty} \delta(\boldsymbol{x}, t - t') \, dt'$$

$$= \boldsymbol{a}(t) \cdot \boldsymbol{\nabla}\delta(\boldsymbol{x}).$$

From (5.13), we see that h has a steady component $\frac{s}{4\pi Dr}$ due to the steady source $s\delta(\boldsymbol{x})$. What we really want is the component of h due to the time-dependent source $-s\boldsymbol{a}(t) \cdot \boldsymbol{\nabla}\delta(\boldsymbol{x}) = -sa_i(t)\partial_i\delta(\boldsymbol{x})$. What can the *derivative* of a δ-function in the source possibly mean? It means "differentiate the point source solution". For our problem, look at $f_i(\boldsymbol{x}, t)$ due to a time-dependent point source at $\boldsymbol{x} = 0$, which satisfies

(5.14) $$\partial_t f_i - D\Delta f_i = -sa_i(t)\delta(\boldsymbol{x}).$$

A particular solution of f_i can be constructed by superposition of diffusion kernels in $\mathbb{R}^3$, analogous to the analysis of the "line source" example in Chapter 4. Differentiating (5.14) with respect to x_i and summing over i, we find that $\partial_i f_i$ satisfies an inhomogeneous diffusion PDE with source $-sa_i(t)\partial_i\delta(\boldsymbol{x})$.

As a simple example, let us take $\boldsymbol{a}(t) = \varepsilon\cos\omega t\,\boldsymbol{e}_1$, so that $f_2 = f_3 = 0$ and f_1 is the real part of the solution to

$$\partial_t f - D\Delta f = -\varepsilon s e^{i\omega t}\delta(\boldsymbol{x}).$$

We seek a separation-of-variables solution,

(5.15) $$f = F(\boldsymbol{x})e^{i\omega t}.$$

The PDE for $F(\boldsymbol{x})$ is

$$\Delta F - \frac{i\omega}{D}F = \frac{\varepsilon s}{D}\delta(\boldsymbol{x}).$$

The δ-function on the right-hand side is equivalent to the "flux condition"

$$\lim_{\varepsilon \to 0} \int_{|\boldsymbol{x}|=\varepsilon} F_r \, da = \frac{s\varepsilon}{D}.$$

The spherically symmetric solution for F which satisfies the flux condition and vanishes as $r \to \infty$ is

$$F = -\frac{\varepsilon s}{4\pi D}\frac{e^{-(1+i)\sqrt{\frac{\omega}{2D}}r}}{r}.$$

The real part of f in (5.15) is

$$f_1 = -\frac{\varepsilon s}{4\pi D}e^{-\sqrt{\frac{\omega}{2D}}r}\cos\left(\omega t - \sqrt{\frac{\omega}{2D}}r\right).$$

The time-periodic component of the h-field due to the source's position oscillating along the e_1-axis is the derivative $\partial_1 f_1 = \partial_r f_1 r_x = \partial_r f_1 e_r \cdot e_1$.

Problem 5.3 (A primer on polarized media). The electric potential field $\varphi(x)$ induced by a charge density $\rho(x)$, x in $\mathbb{R}^3$, satisfies *Poisson's equation*

$$\Delta\varphi = -4\pi\rho.$$

We'll assume the support of ρ is a bounded region; then there is a unique solution of Poisson's equation which vanishes as $r := |x| \to \infty$.

a) Determine the spherically symmetric potential due to a point charge q at the origin, with $\rho(x) = q\delta(x)$.

b) A *dipole* is a pair of opposite and equal charges q and $-q$ separated by some displacement a from the negative to the positive charge. Compute the potential due to the dipole with charges q and $-q$ at positions $\frac{a}{2}$ and $-\frac{a}{2}$. Compute the *point dipole field* φ, which is the leading approximation to the dipole field as $\frac{r}{|a|} \to \infty$. Show that the formal charge density associated with a point dipole is $-p \cdot \nabla\delta(x)$, where $p := qa$ is its *dipole moment*.

c) Next, we look at a configuration of charge made up of little dipoles, as visualized in Figure 5.4.

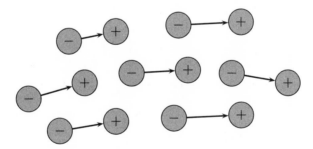

Figure 5.4.

On a macroscopic scale much larger than the individual dipoles, we idealize this system as a "polar medium", a continuum made up of infinitesimal point dipoles. The continuum is characterized by a *polarization vector* field $p(x)$: The effective charge density $\rho(x)$ is the formal superposition of point dipole charge densities:

(5.3-1) $$\rho(x) = -\int_{\mathbb{R}^3} p(x') \cdot \nabla\delta(x - x')\, dx'.$$

Show that $\rho(x) = -(\nabla \cdot p)(x)$.

d) Suppose we have a polar medium in a bounded region R with polarization $\boldsymbol{p}(\boldsymbol{x})$, surrounded by vacuum ($\boldsymbol{p} \equiv 0$ in vacuum). Show that the associated potential field $\varphi(\boldsymbol{x})$ satisfies the jump condition

$$[\varphi_n] = -4\pi \boldsymbol{p} \cdot \boldsymbol{n}$$

on ∂R. (The jump in φ_n represents an effective surface charge density $\boldsymbol{p} \cdot \boldsymbol{n}$ on ∂R.) Find the electric potential due to a sphere of radius a with uniform polarization $\boldsymbol{p} = p\boldsymbol{e}_3$ inside.

Solution.

a) The spherically symmetric harmonic functions which vanish as $r \to \infty$ are $\varphi = \frac{C}{r}$, where C is a constant. Integrating $\Delta\varphi = -4\pi q\delta(\boldsymbol{x})$ about a ball of radius ε about the origin, we get

$$\int_{r=\varepsilon} \varphi_r \, da = -4\pi q.$$

The left-hand side is $-\frac{C}{\varepsilon^2}(4\pi\varepsilon^2) = -4\pi C$, so $C = q$ and $\varphi = \frac{q}{r}$. For future reference, we note that the "unit charge field" $\frac{1}{r}$ satisfies

(5.3-2)
$$\Delta\left(\frac{1}{r}\right) = -4\pi\delta(\boldsymbol{x}).$$

b) The dipole potential is

(5.3-3)
$$\frac{q}{|\boldsymbol{x} - \frac{\boldsymbol{a}}{2}|} - \frac{q}{|\boldsymbol{x} + \frac{\boldsymbol{a}}{2}|} = \frac{q}{r}\left\{\frac{1}{|\boldsymbol{e}_r - \frac{\boldsymbol{a}}{2r}|} - \frac{1}{|\boldsymbol{e}_r + \frac{\boldsymbol{a}}{2r}|}\right\}.$$

Here, we have used $\boldsymbol{x} = r\boldsymbol{e}_r$. In the limit $\frac{a}{r} \to 0$ (where $a := |\boldsymbol{a}|$),

$$\frac{1}{|\boldsymbol{e}_r - \frac{\boldsymbol{a}}{2r}|} = \left\{\left(\boldsymbol{e}_r - \frac{\boldsymbol{a}}{2r}\right)\cdot\left(\boldsymbol{e}_r - \frac{\boldsymbol{a}}{2r}\right)\right\}^{-\frac{1}{2}}$$

$$= \left\{1 - \frac{\boldsymbol{a}\cdot\boldsymbol{e}_r}{r} + \frac{a^2}{4r^2}\right\}^{-\frac{1}{2}}$$

$$= 1 + \frac{\boldsymbol{a}\cdot\boldsymbol{e}_r}{2r} + O(\frac{a^2}{r^2}).$$

We get the second term in the right-hand side of (5.3-3) by replacing $\boldsymbol{a}$ with $-\boldsymbol{a}$. In summary, the approximation to the dipole field (5.3-3) is $q\frac{\boldsymbol{a}\cdot\boldsymbol{e}_r}{r^2} + O(\frac{a^2}{r^2})$ and the *point dipole* field is

(5.3-4)
$$\varphi := q\frac{\boldsymbol{a}\cdot\boldsymbol{e}_r}{r^2}.$$

To determine the effective charge density associated with the point dipole, write (5.3-4) as

$$\varphi = -\boldsymbol{p} \cdot \boldsymbol{\nabla}\left(\frac{1}{r}\right),$$

where $\boldsymbol{p} := q\boldsymbol{a}$ is the dipole moment. Now take the directional derivative $-\boldsymbol{p} \cdot \boldsymbol{\nabla}$ of the "unit charge identity" (5.3-2) to get

$$\Delta\left(-\boldsymbol{p} \cdot \boldsymbol{\nabla}\left(\frac{1}{r}\right)\right) = 4\pi\boldsymbol{p} \cdot \boldsymbol{\nabla}\delta(\boldsymbol{x});$$

thus the charge density associated with a point dipole is

$$\rho(\boldsymbol{x}) = -\boldsymbol{p} \cdot \boldsymbol{\nabla}\delta(\boldsymbol{x}).$$

Another "derivation" of this point dipole charge density is based on "Taylor expansion" of $q\delta(\boldsymbol{x} - \frac{\boldsymbol{a}}{2}) - q\delta(\boldsymbol{x} + \frac{\boldsymbol{a}}{2})$ with respect to $\boldsymbol{a}$.

c) Given any point in $\mathbb{R}^3$, we can put the origin $\boldsymbol{x} = 0$ there. Hence, it is sufficient to show that $\rho(0) = -(\boldsymbol{\nabla} \cdot \boldsymbol{p})(0)$. In (5.3-1) set $\boldsymbol{x} = 0$ to obtain

$$\rho(0) = -\int_{\mathbb{R}^3} \boldsymbol{p}(\boldsymbol{x}') \cdot \boldsymbol{\nabla}\delta(-\boldsymbol{x}')\,d\boldsymbol{x}' = -\int_R \boldsymbol{p}(-\boldsymbol{x}') \cdot \boldsymbol{\nabla}\delta(\boldsymbol{x}')\,d\boldsymbol{x}'.$$

In the last equality, we changed the variable of integration to $-\boldsymbol{x}'$, and since $\boldsymbol{\nabla}\delta(\boldsymbol{x}') = 0$ for $\boldsymbol{x}' \neq 0$, we replaced $\mathbb{R}^3$ by a bounded region R with $\boldsymbol{x}' = 0$ in its interior. Using the divergence theorem and the fact that $\delta(\boldsymbol{x}') = 0$ for $\boldsymbol{x}' \neq 0$, we have

$$\int_R \boldsymbol{p}(-\boldsymbol{x}) \cdot \boldsymbol{\nabla}\delta(\boldsymbol{x})\,d\boldsymbol{x} = \int_R \delta(\boldsymbol{x})(\boldsymbol{\nabla} \cdot \boldsymbol{p})(-\boldsymbol{x})\,d\boldsymbol{x} = (\boldsymbol{\nabla} \cdot \boldsymbol{p})(0),$$

and hence $\rho(0) = -(\boldsymbol{\nabla} \cdot \boldsymbol{p})(0)$.

d) With $\rho(\boldsymbol{x}) = -\boldsymbol{\nabla} \cdot \boldsymbol{p}$, Poisson's equation reads

(5.3-5) $$\Delta\varphi = 4\pi\boldsymbol{\nabla} \cdot \boldsymbol{p}.$$

Let Q be a thin "shell" about ∂R. Integrating (5.3-5) over Q, using the divergence theorem and letting the shell thickness vanish, we get

$$\int_{\partial R} [\varphi_n]\,da = 4\pi\int_{\partial R} [\boldsymbol{p} \cdot \boldsymbol{n}]\,da.$$

Since $\boldsymbol{p} \equiv 0$ outside of R, this reduces to

$$\int_{\partial R} [\varphi_n]\,da = -4\pi\int_{\partial R} \boldsymbol{p} \cdot \boldsymbol{n}\,da,$$

where $\boldsymbol{p}$ represents the surface values of $\boldsymbol{p}(\boldsymbol{x})$ "just inside" ∂R. Hence, the boundary condition is

(5.3-6) $$[\varphi_n] = -4\pi\boldsymbol{p} \cdot \boldsymbol{n}$$

on ∂R.

For the sphere of radius a with uniform polarization $\boldsymbol{p} = p\boldsymbol{e}_3$ inside, $\nabla \cdot \boldsymbol{p} = 0$ inside, so φ is harmonic *inside* as well as outside. The boundary condition (5.3-6) reads

(5.3-7) $$[\varphi_r] = -4\pi p \boldsymbol{e}_3 \cdot \boldsymbol{e}_r.$$

The structure of φ is very simple:

$$\varphi = \left[\begin{array}{ll} C\dfrac{x_3}{a} = C\dfrac{r}{a}\boldsymbol{e}_3 \cdot \boldsymbol{e}_r & \text{in } r < a, \\[2em] C\dfrac{a^2}{r^2}\boldsymbol{e}_3 \cdot \boldsymbol{e}_r & \text{in } r > a, \end{array} \right.$$

where C is a constant. Harmonicity is obvious. φ is linear in $r < a$, and the $r > a$ field is that of a point dipole, with dipole moment $Ca^2\boldsymbol{e}_3$. The proposed φ is obviously continuous across $r = a$. The value of C, $C = \frac{4\pi}{3}ap$, follows from the boundary condition (5.3-7). Hence, the dipole moment of the $r > a$ point dipole field is

$$\boldsymbol{p} = Ca^2\boldsymbol{e}_3 = \frac{4\pi}{3}a^3 p\boldsymbol{e}_3.$$

Evidently, $\boldsymbol{p}$ is the sum of all the dipole moments inside the sphere.

Problem 5.4 (Exit times for Brownian motion).   Brownian particles are released at time 0 from position $\boldsymbol{X}$ inside a bounded region R of $\mathbb{R}^3$. They are absorbed on ∂R. $g(\boldsymbol{x}, \boldsymbol{X}, t)$ denotes the particle distribution as a function of $\boldsymbol{x}$ in R and time $t > 0$, parametrized by $\boldsymbol{X}$ in R. The distribution g satisfies the diffusion PDE for $\boldsymbol{x}$ in R, the absorbing boundary condition $g = 0$ for $\boldsymbol{x}$ on ∂R, and the initial condition $g(\boldsymbol{x}, \boldsymbol{X}, 0) = \delta(\boldsymbol{x} - \boldsymbol{X})$.

a) Explain why

(5.4-1) $$T(\boldsymbol{X}) := \int_0^\infty f(\boldsymbol{X}, t)\, dt$$

with

(5.4-2) $$f(\boldsymbol{X}, t) := \int_R g(\boldsymbol{x}, \boldsymbol{X}, t)\, d\boldsymbol{x}$$

is the average exit time.

b) Carry out a formal calculation to show that $D(\Delta T)(\boldsymbol{X}) = -1$ for $\boldsymbol{X}$ in R.

c) The plausible boundary condition on $T(\boldsymbol{X})$ is $T = 0$ for $\boldsymbol{X}$ on ∂R. What is the exit time function $T(\boldsymbol{X})$ for $\boldsymbol{X} = (X, Y)$ inside the ellipse $\frac{X^2}{a^2} + \frac{Y^2}{b^2} = 1$ in $\mathbb{R}^2$? In particular, what is the average exit time from the center $(0, 0)$?

Solution.

a) $f(\boldsymbol{X}, t)$ in (5.4-2) is the fraction of particles still in R at time $t > 0$. The fraction of particles which exit between times $t_1 > 0$ and $t_2 > t_1$ is

$$f(\boldsymbol{X}, t_1) - f(\boldsymbol{X}, t_2) = - \int_{t_1}^{t_2} f_t(\boldsymbol{X}, t)\, dt,$$

so $-f_t(\boldsymbol{X}, t)$ as a function of $t > 0$ is the *distribution* of exit times. The average exit time is

$$(5.4\text{-}3) \qquad - \int_0^\infty t f_t(\boldsymbol{X}, t)\, dt = - [t f(\boldsymbol{X}, t)]_0^\infty + \int_0^\infty f(\boldsymbol{X}, t)\, dt.$$

We have $f(\boldsymbol{X}, 0) = 1$, and we expect $f(\boldsymbol{X}, t)$ to decay to zero *exponentially* as $t \to \infty$, so that $t f(\boldsymbol{X}, t) \to 0$ as $t \to \infty$. What is left of (5.4-3) is $\int_0^\infty f(\boldsymbol{X}, t)\, dt$, and $T(\boldsymbol{X})$ in (5.4-1) is indeed the average exit time.

b) We start with

$$(5.4\text{-}4) \quad (\Delta T)(\boldsymbol{X}) = \int_0^\infty \Delta f(\boldsymbol{X}, t)\, dt = \int_0^\infty \left(\int_R \Delta_X g(\boldsymbol{x}, \boldsymbol{X}, t)\, d\boldsymbol{x} \right) dt.$$

The notation $\Delta_X g(\boldsymbol{x}, \boldsymbol{X}, t)$ means the Laplacian of g with respect to $\boldsymbol{X}$. $\Delta_X g(\boldsymbol{x}, \boldsymbol{X}, t)$ as a function of $\boldsymbol{x}$ and t satisfies the diffusion PDE (take Δ_X of $g_t = D\Delta_x g$), vanishes for $\boldsymbol{x}$ on ∂R, and has initial condition

$$\Delta_X g(\boldsymbol{x}, \boldsymbol{X}, 0) = \Delta_X \delta(\boldsymbol{x} - \boldsymbol{X}) = \Delta_x \delta(\boldsymbol{x} - \boldsymbol{X}).$$

The main point is the last equality, which trades Δ_X for Δ_x. It is now evident that

$$\Delta_X g(\boldsymbol{x}, \boldsymbol{X}, t) = \Delta_x g(\boldsymbol{x}, \boldsymbol{X}, t) = \frac{1}{D} g_t(\boldsymbol{x}, \boldsymbol{X}, t)$$

for $\boldsymbol{x}$, $\boldsymbol{X}$ in R and $t > 0$. Hence, (5.4-4) becomes

$$(\Delta T)(\boldsymbol{X}) = \frac{1}{D} \int_0^\infty \int_R g_t(\boldsymbol{x}, \boldsymbol{X}, t)\, d\boldsymbol{x}\, dt$$

$$= \frac{1}{D} \int_R \{ g(\boldsymbol{x}, \boldsymbol{X}, \infty) - g(\boldsymbol{x}, \boldsymbol{X}, 0) \}\, d\boldsymbol{x} = -\frac{1}{D};$$

so finally,

$$D(\Delta T)(\boldsymbol{X}) = -1$$

for $\boldsymbol{X}$ in R.

c) We propose

$$T(X, Y) = T_0 \left(1 - \frac{X^2}{a^2} - \frac{Y^2}{b^2} \right),$$

where T_0 is a constant. We compute

$$-1 = D\Delta T = -2DT_0\left(\frac{1}{a^2} + \frac{1}{b^2}\right),$$

and hence

$$T_0 = \frac{1}{2D}\frac{a^2b^2}{a^2 + b^2}$$

is the average exit time from $(0,0)$.

δ-**sequences.** The role of arbitrary "test functions" $f(x)$ in the definition (5.3) of a δ-sequence is motivated by basic problems in analysis which give rise to δ-sequences in the first place. The long-familiar diffusion kernel

$$(5.16) \qquad\qquad g(x,t) = \frac{1}{2\sqrt{\pi Dt}}e^{-\frac{x^2}{4Dt}}$$

is a simple example. Recall our proposed solution $c(x,t)$ to the diffusion PDE with given initial condition $c(x,0)$,

$$(5.17) \qquad\qquad c(x,t) = \int_{-\infty}^{\infty} c(x',0)g(x - x', t)\, dx'.$$

The limit of the right-hand side as $t \to 0$ should recover the initial value $c(x,0)$, and this is so because (5.16) is a δ-sequence in the sense of definition (5.3), with $2\sqrt{Dt}$ in the role of ε,

$$(5.18) \qquad\qquad \delta(x,\varepsilon) = \frac{e^{-\frac{x^2}{\varepsilon^2}}}{\sqrt{\pi}\varepsilon}.$$

For bounded continuous $f(x)$, we wish to show that

$$(5.19) \qquad\qquad \int_{-\infty}^{\infty} f(x)\frac{e^{-\frac{x^2}{\varepsilon^2}}}{\sqrt{\pi}\varepsilon}\, dx$$

converges to $f(0)$ as $\varepsilon \to 0$. We start with a preliminary estimate that shortens the analysis: Let F be an upper bound on $|f(x)|$; then for $\mu > 0$,

$$(5.20) \qquad\qquad \left|\int_{|x|>\mu} f(x)\frac{e^{-\frac{x^2}{\varepsilon^2}}}{\sqrt{\pi}\varepsilon}\, dx\right| \leq \frac{F}{\sqrt{\pi}}\frac{\varepsilon}{\mu}.$$

The above estimate is obtained as follows:

$$\left| \int_{|x|>\mu} f(x) \frac{e^{-\frac{x^2}{\varepsilon^2}}}{\sqrt{\pi}\varepsilon}\, dx \right| \leq 2F \int_{\mu}^{\infty} \frac{e^{-\frac{x^2}{\varepsilon^2}}}{\sqrt{\pi}\varepsilon}\, dx$$

$$= \frac{2F}{\sqrt{\pi}\varepsilon} e^{-\frac{\mu^2}{\varepsilon^2}} \int_{\mu}^{\infty} e^{-\frac{(x+\mu)(x-\mu)}{\varepsilon^2}}\, dx$$

$$\leq \frac{2F}{\sqrt{\pi}\varepsilon} e^{-\frac{\mu^2}{\varepsilon^2}} \int_{\mu}^{\infty} e^{-\frac{2\mu(x-\mu)}{\varepsilon^2}}\, dx$$

$$= \frac{F}{\sqrt{\pi}} \frac{\varepsilon}{\mu} e^{-\frac{\mu^2}{\varepsilon^2}}.$$

Let $\mu = \mu(\varepsilon)$ be any "gauge function" with $\varepsilon \ll \mu(\varepsilon) \ll 1$, meaning that $\mu(\varepsilon) \to 0$ and $\frac{\varepsilon}{\mu(\varepsilon)} \to 0$ as $\varepsilon \to 0$. We separate the integral (5.19) into $|x| < \mu$ and $|x| > \mu$ components. The $|x| > \mu$ component vanishes as $\varepsilon \to 0$ by (5.20). Hence, it is sufficient to show that the $|x| < \mu$ component converges to $f(0)$ as $\varepsilon \to 0$, that is,

$$\int_{-\mu}^{\mu} f(x) \frac{e^{-\frac{x^2}{\varepsilon^2}}}{\sqrt{\pi}\varepsilon}\, dx \to f(0)$$

as $\varepsilon \to 0$. We write the integral on the left-hand side as

$$\int_{-\infty}^{\infty} f(0) \frac{e^{-\frac{x^2}{\varepsilon^2}}}{\sqrt{\pi}\varepsilon}\, dx - \int_{|x|>\mu} f(0) \frac{e^{-\frac{x^2}{\varepsilon^2}}}{\sqrt{\pi}\varepsilon}\, dx + \int_{-\mu}^{\mu} (f(x) - f(0)) \frac{e^{-\frac{x^2}{\varepsilon^2}}}{\sqrt{\pi}\varepsilon}\, dx.$$

The first integral is $f(0)$. The second vanishes as $\varepsilon \to 0$ by (5.20) and the assumption that $\frac{\varepsilon}{\mu(\varepsilon)} \to 0$ as $\varepsilon \to 0$. The third is bounded in absolute value by

$$\max_{|x|<\mu} |f(x) - f(0)|,$$

and this vanishes as $\varepsilon \to 0$ by the continuity of f and the assumption that $\mu(\varepsilon) \to 0$ as $\varepsilon \to 0$.

The diffusion kernel is a very "simple" δ-sequence: The convergence of $\delta(x, \varepsilon)$ to zero as $\varepsilon \to 0$ with $x \neq 0$ fixed is so strong that, in essence, "its support is shrinking to zero" as $\varepsilon \to 0$. As long as $\int_{-\infty}^{\infty} \delta(x, \varepsilon)\, dx \to 1$ as $\varepsilon \to 0^+$, such "simple" $\delta(x, \varepsilon)$ are bound to be δ-sequences in the sense of the definition (5.3). The proof of (5.3) as an identity for such "simple" $\delta(x, \varepsilon)$

is useful, especially since *convolution* integrals of the form

$$(5.21) \quad (f * \delta)(\boldsymbol{x}) := \int_{\mathbb{R}^n} f(\boldsymbol{x}')\delta(\boldsymbol{x} - \boldsymbol{x}', \varepsilon)\, d\boldsymbol{x}' = \int_{\mathbb{R}^n} f(\boldsymbol{x} - \boldsymbol{x}')\delta(\boldsymbol{x}', \varepsilon)\, d\boldsymbol{x}'$$

routinely arise in analysis, and establishing the convergence of $(f * \delta)(\boldsymbol{x})$ to $f(\boldsymbol{x})$ as $\varepsilon \to 0$ is a key point. However, if the "simple" $\delta(\boldsymbol{x}, \varepsilon)$ are the only kind we ever see, then insistence on "test functions" in the definition (5.3) looks superfluous. But they are not and it isn't.

Problem 5.5 (Green's identity and inhomogeneous diffusion PDEs).

a) Let $u(x, t)$ and $v(x, t)$ be analytic functions of x and t and let R be a bounded region of the (x, t) plane. Show that

(5.5-1)
$$\int_R \{v(u_t - Du_{xx}) + u(v_t + Dv_{xx})\}\, dx\, dt = \int_{\partial R} -uv\, dx + D(uv_x - vu_x)\, dt.$$

b) Let $c(x, t)$ be a solution of the inhomogeneous diffusion PDE

$$(5.5\text{-}2) \qquad\qquad c_t - Dc_{xx} = s(x, t),$$

and let $h(x, t)$ be a solution of the "backwards" diffusion PDE

$$(5.5\text{-}3) \qquad\qquad h_t = -Dh_{xx}$$

in the region R in Part a. Show that

$$(5.5\text{-}4) \qquad \int_R hs\, dx\, dt = \int_{\partial R} -ch\, dx + D(ch_x - hc_x)\, dt.$$

c) Assume that the source $s(x, t)$ in (5.5-2) is a bounded continuous function whose support lies in the half space $t > t_m$, a given value, and that $c(x, t)$ satisfies the "causal" initial condition $c = 0$ for $t \le t_m$. Further, assume c and c_x are bounded. Show that for any given X, T, and $\varepsilon > 0$,

(5.5-5)
$$\int_{-\infty}^{T-\varepsilon} \int_{-\infty}^{\infty} g(x - X, T - t)s(x, t)\, dx\, dt$$
$$= \int_{-\infty}^{\infty} g(x, \varepsilon)c(X + x, T - \varepsilon)\, dx,$$

where $g(x, t)$ is the diffusion kernel.

d) Derive an integral representation of $c(X, T)$ by taking the $\varepsilon \to 0$ limit of (5.5-5).

Solution.

a) The integrand of the left-hand side of (5.5-1) can be rewritten as

$$D(uv_x - vu_x)_x - (-uv)_t.$$

By Green's theorem,

$$\int_R \{D(uv_x - vu_x)_x - (-uv)_t\}\, dx\, dt = \int_{\partial R} -uv\, dx + D(uv_x - vu_x)\, dt.$$

b) In (5.5-1), take $u = c$ and $v = h$ to get

$$\int_R \{h(c_t - Dc_{xx}) + c(h_t + Dh_{xx})\}\, dx\, dt = \int_R -ch\, dx + D(ch_x - hc_x)\, dt.$$

By (5.5-2) and (5.5-3), the left-hand side is $\int_R hs\, dx\, dt$.

c) Take $h(x,t) = g(x - X, T - t)$, a solution of the backwards diffusion PDE in $t < T$. Take R to be the rectangle with $t_m < t < T - \varepsilon$ and $-L < x - X < L$, as depicted in Figure 5.5.

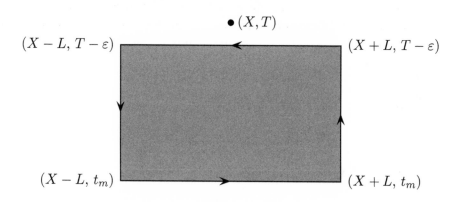

Figure 5.5.

The left-hand side of (5.5-4) is

$$\int_R s(x,t)g(x - X, T - t)\, dx\, dt.$$

Due to the decay of g as $|x - X| \to \infty$, the boundedness of s, and the assumption that $s = 0$ for $t < t_m$, this converges to

(5.5-6) $$\int_{-\infty}^{\infty} \int_{-\infty}^{T-\varepsilon} s(x,t)g(x - X, T - t)\, dx\, dt$$

as $L \to \infty$. Now look at the line integral on the right-hand side of (5.5-4). The contribution from $t = t_m$ vanishes since $c = 0$ on $t = t_m$. The contributions from $x = X + L$ and $x = X - L$ vanish as $L \to \infty$ due to strong decay of h and h_x and boundedness of c and c_x. Hence, in the limit $L \to \infty$, only the contribution from $t = T - \varepsilon$ survives, and this is

$$(5.5\text{-}7) \quad \int_{\infty}^{-\infty} -c(x, T-\varepsilon)g(x-X, T-\varepsilon)\, dx = \int_{-\infty}^{\infty} c(X+x, T-\varepsilon)g(x, \varepsilon)\, dx.$$

Equating (5.5-6) and (5.5-7) gives (5.5-5).

d) Due to the δ-sequence character of $g(x, \varepsilon)$, the right-hand side of (5.5-5) converges to $c(X, T)$ as $\varepsilon \to 0$. Hence, the $\varepsilon \to 0$ limit of (5.5-5) is

$$c(X, T) = \int_{-\infty}^{T} \int_{-\infty}^{\infty} s(x, t)\, g(x - X, T - t)\, dx\, dt.$$

"Oscillatory" δ-sequences. We begin with the famous δ-sequence (Churchill & Brown [6]) which is at the heart of Fourier series. Let $f(x)$ be a continuous complex 2π-periodic function of x in $\mathbb{R}$. The trigonometric sum

$$(5.22) \qquad\qquad S_N(x) = \sum_{-N}^{N} \hat{f}_n e^{inx},$$

which is the least-squares approximation to $f(x)$, has

$$(5.23) \qquad\qquad \hat{f}_n = \frac{1}{2\pi} \int_{-\pi}^{\pi} f(x')e^{-inx'}\, dx'.$$

Specifically, the choices (5.23) of coefficients $\hat{f}_n$ minimize

$$\int_{-\pi}^{\pi} |f(x) - S_N(x)|^2\, dx.$$

We investigate the convergence of $S_N(x)$ to $f(x)$ as $N \to \infty$. First we substitute the coefficients (5.23) into (5.22) and interchange the order of summation and integration:

$$S_N(x) = \int_{-\pi}^{\pi} f(x') \left\{ \frac{1}{2\pi} \sum_{-N}^{N} e^{in(x-x')} \right\} dx'$$

$$= \int_{-\pi}^{\pi} f(x - x') \left\{ \frac{1}{2\pi} \sum_{-N}^{N} e^{inx'} \right\} dx'.$$

The geometric series in the right-hand side can be summed explicitly, giving

$$(5.24) \qquad\qquad S_N(x) = \int_{-\pi}^{\pi} f(x - x') \frac{1}{2\pi} \frac{\sin(N + \frac{1}{2})x'}{\sin \frac{x'}{2}}\, dx'.$$

The essential idea of the convergence proof is that

(5.25)
$$\delta_N(x) := \frac{\sin(N + \frac{1}{2})x}{\sin \frac{x}{2}}$$

is a discrete δ-sequence on $[-\pi, \pi]$, that is,

(5.26)
$$\lim_{N\to\infty} \int_{-\pi}^{\pi} \varphi(x)\delta_N(x)\, dx = \varphi(0)$$

for any continuous function $\varphi(x)$ on $[-\pi, \pi]$. The convergence of $S_N(x)$ in (5.24) to $f(x)$ as $N \to \infty$ is an immediate consequence.

The (technical) proof of (5.26) is a standard result in Fourier analysis. Here, we want to emphasize that $\delta_N(x)$ has no pointwise limit as $N \to \infty$ with x fixed. Figure 5.6 presents the graph of $\delta_N(x)$ for $N = 128$. Away

$$\delta_{128}(x)$$

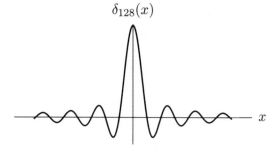

Figure 5.6.

from $x = 0$, we see rapid oscillations with period $\frac{2\pi}{N+1/2}$ whose amplitude $\frac{1}{\sin(x/2)}$ is independent of N. In the integral (5.26) there is oscillation-induced cancellation of contributions away from $x = 0$, and the main contribution comes from a neighborhood of the origin $x = 0$ with width on the order of $\frac{1}{N}$. In particular, for $\varphi(x) \equiv 1$, the integral (5.26) reduces to

$$\int_{-\pi}^{\pi} \delta_N(x)\, dx = \sum_{-N}^{N} \frac{1}{2\pi} \int_{-\pi}^{\pi} e^{inx}\, dx = 1,$$

and the contribution from the "spike" $-\frac{\pi}{N+1/2} < x < \frac{\pi}{N+1/2}$, namely

$$\int_{-\frac{\pi}{N+1/2}}^{\frac{\pi}{N+1/2}} \delta_N(x)\, dx,$$

asymptotes to

$$\frac{2}{\pi} \int_0^{\pi} \frac{\sin x}{x}\, dx \approx 1.17898$$

as $N \to \infty$.

Problem 5.6 (Cancellation by oscillation).

a) Let $p(x)$ be 2π-periodic in x, with $\int_{-\pi}^{\pi} p(x)\,dx = 0$. $\theta(x)$ is twice continuously differentiable in $[a, b]$, with $\theta'(x) > h$, a positive constant in $[a, b]$. $\varphi(x)$ is continuously differentiable in $[a, b]$. Show that

$$(5.6\text{-}1) \qquad \lim_{\varepsilon \to 0} \int_a^b \varphi(x) p\left(\frac{\theta(x)}{\varepsilon}\right) dx = 0.$$

b) Prove the δ-sequence identity (5.26) for twice continuously differentiable $\varphi(x)$.

Solution.

a) Let $P(x)$ be an antiderivative of $p(x)$. Because $\int_{-\pi}^{\pi} p(x)\,dx = 0$, $P(x)$ is 2π-periodic too. We calculate

$$\int_a^b \varphi(x) p\left(\frac{\theta(x)}{\varepsilon}\right) dx = \varepsilon \int_a^b \frac{\varphi}{\theta'} \frac{\theta'}{\varepsilon} P'\left(\frac{\theta}{\varepsilon}\right) dx$$

$$(5.6\text{-}2) \qquad\qquad = \varepsilon \int_a^b \frac{\varphi}{\theta'} \left(P\left(\frac{\theta}{\varepsilon}\right)\right)' dx$$

$$= \varepsilon \left[\frac{\varphi}{\theta'} P\left(\frac{\theta}{\varepsilon}\right)\right]_a^b - \varepsilon \int_a^b \left(\frac{\varphi}{\theta'}\right)' P\left(\frac{\theta}{\varepsilon}\right) dx.$$

Let $P_m = \max|P|$. Since $\theta' > h$, $\gamma(x) := \frac{\varphi(x)}{\theta'(x)}$ is continuously differentiable in $[a, b]$. Let γ_m and γ'_m be the maxima of $|\gamma(x)|$ and $|\gamma'(x)|$ in $[a, b]$. It follows from (5.6-2) that

$$\left|\int_a^b \varphi p\left(\frac{\theta}{\varepsilon}\right) dx\right| \leq \varepsilon P_m (2\gamma_m + \gamma'_m(b - a)),$$

and (5.6-1) is immediate.

b) First, write

$$(5.6\text{-}3) \qquad \int_{-\pi}^{\pi} \varphi(x) \frac{\sin(N + \frac{1}{2})x}{\sin\frac{x}{2}}\,dx$$

$$= \varphi(0) + \int_{-\pi}^{\pi} \frac{\varphi(x) - \varphi(0)}{\sin\frac{x}{2}} \sin(N + \frac{1}{2})x\,dx.$$

Here, we have used

$$\int_{-\pi}^{\pi} \frac{\sin(N + \frac{1}{2})x}{\sin\frac{x}{2}}\,dx = 1.$$

In the right-hand side of (5.6-3), notice that

$$\gamma(x) := \begin{bmatrix} \frac{\varphi(x)-\varphi(0)}{\sin \frac{x}{2}}, & x \neq 0, \\ \\ 2\varphi'(0), & x = 0 \end{bmatrix}$$

is continuously differentiable in $[-\pi, \pi]$. By the calculation in Part a, the integral in the right-hand side of (5.6-3) is bounded above by a positive constant times $\frac{1}{N+1/2}$. Hence,

$$\lim_{N \to \infty} \int_{-\pi}^{\pi} \varphi(x) \frac{\sin(N+\frac{1}{2})x}{\sin \frac{x}{2}} \, dx = \varphi(0).$$

Linear PDEs routinely generate examples of "oscillatory δ-sequences". For instance, consider the initial value problem

(5.27) $$c_t = \frac{1}{3}c_{xxx}, \quad \text{all } x, \text{ all } t > 0,$$

(5.28) $$c(x, 0) \text{ given.}$$

Suppose we find a solution $g(x, t)$ of the PDE (5.27) which is a δ-sequence with t in the role of ε. Then the solution of the initial value problem has integral representation (5.17) with the new $g(x, t)$.

What is this g? The PDE (5.27) and the integral condition, which states that $\int_{-\infty}^{\infty} g(x, t) \, dx = 1$ for all t, both have a scaling symmetry: If $g(x, t)$ satisfies the PDE and integral condition, then so does $G(x, t) := \frac{1}{a}g(ax, a^3 t)$, for any positive constant a. It follows that a solution that is invariant under these scalings, with $G(x, t) = g(x, t)$, takes the self-similar form

(5.29) $$g(x, t) = t^{-1/3}\gamma(\zeta := xt^{-1/3}).$$

Substituting (5.29) into (5.27) gives an ODE for $\gamma(\zeta)$,

$$(\gamma'' + \zeta\gamma)' = 0$$

for all ζ. Hence, $\gamma'' + \zeta\gamma = k$, a constant independent of ζ. If $k \neq 0$, the solution for γ which asymptotes to zero as $\zeta \to -\infty$ would have $\zeta\gamma \to k$ as $\zeta \to -\infty$. But then γ would not be integrable at $\zeta = -\infty$, and we can't enforce $\int_{-\infty}^{\infty} \gamma(\zeta) \, d\zeta = 1$, which in turn implies $\int_{-\infty}^{\infty} g(x, t) \, dx = 1$. Hence, take $k = 0$, so that

$$\gamma'' + \zeta\gamma = 0.$$

The solution of this ODE on $-\infty < \zeta < \infty$ with $\int_{-\infty}^{\infty} \gamma \, d\zeta = 1$ is

$$\gamma(\zeta) = \text{Ai}(-\zeta),$$

where Ai(s) is a special function (the "Airy function"). The graph of $\gamma(\zeta)$ is shown in Figure 5.7. We propose that the δ-sequence solution of the PDE (5.27) is

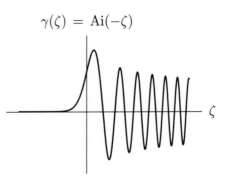

$$\gamma(\zeta) = \mathrm{Ai}(-\zeta)$$

(5.30) $g(x,t) = t^{-1/3}\mathrm{Ai}(-xt^{-1/3}).$

The solution $g(x,t)$ rapidly converges to zero as $t \to 0$ with $x < 0$ fixed. However, there is an oscillatory "tail" in $x > 0$, and it is evident that the am-

Figure 5.7.

plitude and frequency of these $x > 0$ oscillations grow as $t \to 0$ with $x > 0$ fixed. In fact, the asymptotic behavior of $\gamma(\zeta) = \mathrm{Ai}(-\zeta)$ as $\zeta \to +\infty$ is

(5.31) $$\gamma(\zeta) \sim \frac{1}{\sqrt{\pi}} |\zeta|^{-1/4} \sin\left(\frac{2}{3}|\zeta|^{3/2} + \frac{\pi}{4}\right).$$

Here, "$\sim$" means that the relative error of the approximation on the right-hand side goes to zero as $\zeta \to \infty$. It follows from (5.30) and (5.31) that

(5.32) $$g(x,t) \sim \frac{1}{\sqrt{\pi}} (xt)^{-1/4} \sin\left(\frac{2}{3}x^{3/2}t^{-1/2} + \frac{\pi}{4}\right)$$

as $t \to 0$ with $x > 0$ fixed. In this same limit, we see that the amplitude of the oscillations grows like $t^{-1/4}$, and the local frequency $x^{1/2}t^{-1/2}$ grows like $t^{-1/2}$.

So what happens to the convolution integral (5.17) as $t \to 0$? Although the oscillations of $g(x,t)$ grow in amplitude as $t \to 0$ with $x > 0$ fixed, they become so *rapid* that there is a lot of cancellation, and it can be shown that the integral (5.17) converges to $c(x,0)$ anyway.

Problem 5.7 (An oscillatory δ-sequence).

a) Determine a complex similarity solution of the Schrödinger PDE

$$-i\psi_t = \psi_{xx}$$

in $t > 0$, for all x, which has even symmetry in x and satisfies the integral condition

$$\int_{-\infty}^{\infty} \psi \, dx = 1.$$

Explain how $\psi(x,t)$ fails to have a pointwise limit as $t \to 0$ with $x \neq 0$ fixed.

b) We examine the δ-sequence nature of $g(x, \varepsilon) := \psi(x, \frac{\varepsilon^2}{4})$ in the sense of the definition

(5.7-1) $$\lim_{\varepsilon \to 0} \int_{-\infty}^{\infty} \varphi(x) g(x, \varepsilon) \, dx = \varphi(0)$$

for all continuous, bounded $\varphi(x)$. Show that (5.7-1) holds for the more restrictive class of $\varphi(x)$ which are also smooth and have

$$\gamma(x) := \begin{bmatrix} \frac{\varphi(x) - \varphi(0)}{x}, & x \neq 0, \\[2mm] \varphi'(0), & x = 0, \end{bmatrix}$$

with $\gamma'(x)$ absolutely integrable, i.e., $\int_{-\infty}^{\infty} |\gamma'(x)| \, dx < \infty$. (Hint: The additional hypotheses on $\varphi(x)$ are tailored for a calculation that mimics Problem 5.6.)

Solution.

a) The Schrödinger PDE written as $\psi_t = i\psi_{xx}$ looks like the diffusion PDE with $D = i$. So a heuristic answer is to set $D = i$ in the diffusion kernel, giving

(5.7-2) $$\psi(x,t) = \frac{e^{-\frac{x^2}{4it}}}{2\sqrt{\pi i t}} = \frac{e^{i\frac{x^2}{4t}}}{2\sqrt{\pi i t}}.$$

If you actually do what is asked, you would start with

(5.7-3) $$\psi(x,t) = \frac{1}{\sqrt{t}} S\left(\zeta := \frac{x}{\sqrt{t}}\right).$$

The prefactor $\frac{1}{\sqrt{t}}$ ensures that $\int_{-\infty}^{\infty} \psi \, dx$ is independent of t. You would then get an ODE for $S(\zeta)$. The even solution with $\int_{-\infty}^{\infty} S(\zeta) \, d\zeta = 1$ is $S = \frac{e^{i\zeta^2/4}}{2\sqrt{\pi i}}$, and (5.7-3) with this S is (5.7-2). For $t \to 0$ with x bounded away from zero, $\psi(x,t)$ exhibits oscillations of local spatial frequency $\frac{x}{2t}$ and local amplitude $\frac{1}{2\sqrt{\pi t}}$, which blow up *everywhere* as $t \to 0$.

b) Following the first step in the solution of Problem 5.6b,

(5.7-4) $$\int_{-\infty}^{\infty} \varphi(x) g(x,t) \, dx = \varphi(0) + \int_{-\infty}^{\infty} (\varphi(x) - \varphi(0)) g(x,\varepsilon) \, dx.$$

Motivated by the analysis in Problem 5.6a, we write

$$g(x,\varepsilon) = \frac{e^{i\frac{x^2}{\varepsilon^2}}}{\sqrt{\pi i \varepsilon}} = \frac{\varepsilon}{2i\sqrt{\pi i}} \frac{1}{x} \left(e^{i\frac{x^2}{\varepsilon^2}}\right)',$$

and (5.7-4) becomes

$$(5.7\text{-}5) \qquad \varphi(0) + \frac{\varepsilon}{2i\sqrt{\pi i}} \int_{-\infty}^{\infty} \frac{\varphi(x) - \varphi(0)}{x} \left(e^{i\frac{x^2}{\varepsilon^2}} \right)' dx.$$

Here, $\gamma(x) := \frac{\varphi(x) - \varphi(0)}{x}$ is a nice function with a removable singularity at $x = 0$. We do the integration by parts suggested by (5.7-5) and obtain

$$\int_{-\infty}^{\infty} \varphi(x) g(x, \varepsilon)\, dx = \varphi(0) - \frac{\varepsilon}{2i\sqrt{\pi i}} \int_{-\infty}^{\infty} \gamma'(x) e^{i\frac{x^2}{\varepsilon^2}}\, dx.$$

Since the integral on the right-hand side is bounded in absolute value by $\varepsilon \int_{-\infty}^{\infty} |\gamma'(x)|\, dx$, (5.7-1) is immediate.

Guide to bibliography. Recommended references for this chapter are Churchill & Brown [6], Courant & Hilbert volume II [9], Erdélyi [10], Gel'fand & Shilov [14], John [15], and Lighthill [19].

Gel'fand & Shilov [14] present the standard conversion of δ-function source terms in linear ODEs into equivalent jump conditions. The intention of their book is the rigorous theory of generalized functions, but simple "common sense" results can nevertheless be recognized, once the discussion lands on concrete ODEs and PDEs. The notion of δ-sequence and elementary examples thereof are presented in Courant & Hilbert volume II [9], Erdélyi [10], Gel'fand & Shilov [14], and Lighthill [19]. In particular, Lighthill [19] gives a short proof of the δ-sequence identity (5.3) in the one-dimensional case, with the Gaussian δ-sequence. John [15] prefaces his rigorous analysis of solutions to Poisson's equation by a formal calculation, essentially like our "δ-function shorthand explanation of superposition", which is discussed right after equation (5.9). John also rigorously establishes the solution of the diffusion PDE initial value problem based on superposition of diffusion kernels. Churchill & Brown [6] present the classic proof of convergence of Fourier series, based on convolution of a given function with the oscillatory δ-sequence $\delta_N(x)$ in (5.25). Problem 5.6 in our text mimics calculations (including use of the Riemann-Lebesgue lemma) that are part of their proof.

Part 3

Scaling-based reductions in basic fluid mechanics

Chapter 6

Ideal fluid mechanics

Ideal fluid. In a material continuum, the state variables which describe the distribution and motion of a matter are the mass density $\rho(\boldsymbol{x}, t)$ and the velocity field $\boldsymbol{u}(\boldsymbol{x}, t)$ which convects the matter. If $R(t)$ is any material region associated with the flow $\boldsymbol{u}$, conservation of mass is expressed by

$$(6.1) \qquad \frac{d}{dt} \int_{R(t)} \rho \, d\boldsymbol{x} = 0.$$

The nature of the material continuum—as fluid, solid, or other—is specified by the mechanics of momentum and energy transfer between material regions. Exchange of momentum due to short (molecular) range forces is described by the *stress*, or force per unit area acting on the boundary of a material region. In the simplest model of a fluid, called *ideal fluid*, the only stress is due to the *pressure field* $p(\boldsymbol{x}, t)$: The pressure force on fluid in $R(t)$ is $- \int_{\partial R(t)} p \boldsymbol{n} \, da$. The action of the pressure field is visualized in Figure 6.1.

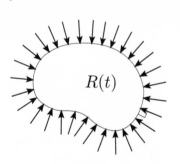

Figure 6.1.

The pressure force on $R(t)$ determines the rate of change of momentum in $R(t)$ according to Newtonian mechanics:

$$(6.2) \qquad \frac{d}{dt} \int_{R(t)} \rho \boldsymbol{u} \, d\boldsymbol{x} = - \int_{\partial R(t)} p \boldsymbol{n} \, da.$$

Energy balance in a fluid is more subtle. In addition to the kinetic energy $\frac{1}{2}|\boldsymbol{u}|^2$ per unit mass associated with the "bulk" velocity $\boldsymbol{u}(\boldsymbol{x}, t)$, there is energy associated with microscopic degrees of freedom: Random thermal motions may be too small to "see" in bulk mass transport, but they can make a significant additive contribution to the kinetic energy per unit mass.

There is also a potential energy per unit mass arising from molecular-range forces. The sum of all energies per unit mass *not* contained in $\frac{1}{2}|\boldsymbol{u}|^2$ is called the *internal energy e*. For fluids in thermal and mechanical equilibrium, with ρ, $\boldsymbol{u}$, p all uniform and constant, there is a functional relation

(6.3) $$e = e(\rho, p),$$

called the *equation of state*. It is a basic assumption of continuum mechanics that the equation of state applies locally in media with nonuniform, nonconstant ρ, $\boldsymbol{u}$, p. In summary, the energy per unit mass is

$$\frac{1}{2}|\boldsymbol{u}|^2 + e(\rho, p)$$

and the total energy in a material region $R(t)$ is

$$\int_{R(t)} \rho \left(\frac{1}{2}|\boldsymbol{u}|^2 + e(\rho, p) \right) \, dx.$$

How does this energy change? The pressure force does work on the matter in $R(t)$ at rate

$$-\int_{R(t)} p\, \boldsymbol{n} \cdot \boldsymbol{u}\, da.$$

Any energy exchange between $R(t)$ and its environment which is not work done by stress is called *heat transfer*. In the ideal fluid model, heat transfer is very simple—there is none. Hence, energy balance in ideal fluid is expressed by

(6.4) $$\frac{d}{dt} \int_{R(t)} \rho \left(\frac{1}{2}|\boldsymbol{u}|^2 + e(\rho, p) \right) \, dx = - \int_{\partial R(t)} p \boldsymbol{u} \cdot \boldsymbol{n}\, da$$

The three integral identities (6.1), (6.2) and (6.4) of mass, momentum, and energy balance together with the equation of state (6.3) completely specify the physics of ideal fluid. Local conservation laws of mass, momentum, and energy readily follow. Applying the transport identity (1.4) to the mass conservation (6.1) gives

(6.5) $$\int_{R(t)} \{\rho_t + \boldsymbol{\nabla} \cdot (\rho \boldsymbol{u})\} \, d\boldsymbol{x} = 0.$$

The ith component of momentum balance (6.2) is

$$\frac{d}{dt} \int_{R(t)} \rho u_i \, d\boldsymbol{x} = - \int_{\partial R(t)} (p\, \boldsymbol{e}_i) \cdot \boldsymbol{n}\, da.$$

Applying the transport identity to the left-hand side and the divergence theorem to the right-hand side gives

$$(6.6) \qquad \int_{R(t)} \left\{ (\rho u_i)_t + \boldsymbol{\nabla} \cdot ((\rho u_i)\boldsymbol{u} + p\,\boldsymbol{e}_i) \right\} d\boldsymbol{x} = 0.$$

Similar application of the transport identity and divergence theorem to the energy identity (6.4) gives

$$(6.7) \qquad \int_{R(t)} \left\{ \left(\frac{\rho}{2}|\boldsymbol{u}|^2 + \rho e \right)_t + \boldsymbol{\nabla} \cdot \left(\left(\frac{\rho}{2}|\boldsymbol{u}|^2 + \rho e \right)\boldsymbol{u} + p\boldsymbol{u} \right) \right\} d\boldsymbol{x} = 0.$$

Since the integral identities (6.5)–(6.7) are to hold for all material regions $R(t)$, we have local conservation laws of mass, momentum and energy:

$$(6.8) \qquad \rho_t + \boldsymbol{\nabla} \cdot (\rho \boldsymbol{u}) = 0,$$

$$(6.9) \qquad (\rho u_i)_t + \boldsymbol{\nabla} \cdot (\rho u_i \boldsymbol{u} + p\,\boldsymbol{e}_i) = 0, \quad i = 1, 2, 3,$$

$$(6.10) \qquad \left(\frac{\rho}{2}|\boldsymbol{u}|^2 + \rho e \right)_t + \boldsymbol{\nabla} \cdot \left\{ \left(\frac{\rho}{2}|\boldsymbol{u}|^2 + \rho e \right)\boldsymbol{u} + p\boldsymbol{u} \right\} = 0.$$

A reformulation in terms of the convective derivative simplifies the physical interpretation of these PDEs:

$$(6.11) \qquad \frac{D\rho}{Dt} + \rho \boldsymbol{\nabla} \cdot \boldsymbol{u} = 0,$$

$$(6.12) \qquad \frac{D}{Dt}(\rho u_i) + (\rho u_i)\boldsymbol{\nabla} \cdot \boldsymbol{u} + \partial_i p = 0, \quad i = 1, 2, 3,$$

$$(6.13) \qquad \frac{D}{Dt}\left(\frac{\rho}{2}|\boldsymbol{u}|^2 + \rho e \right) + \left(\frac{\rho}{2}|\boldsymbol{u}|^2 + \rho e \right)\boldsymbol{\nabla} \cdot \boldsymbol{u} + \boldsymbol{\nabla} \cdot (p\boldsymbol{u}) = 0.$$

From (6.11),

$$(6.14) \qquad \boldsymbol{\nabla} \cdot \boldsymbol{u} = -\frac{1}{\rho}\frac{D\rho}{Dt},$$

and substitution into (6.12) gives

$$\frac{D}{Dt}(\rho u_i) - u_i \frac{D\rho}{Dt} + \partial_i p = 0, \quad i = 1, 2, 3.$$

Apply the product rule to the first term and rearrange what is left. We get

$$\frac{Du_i}{Dt} = -\frac{1}{\rho}\partial_i p, \quad i = 1, 2, 3.$$

These equations are components of a vector PDE

$$(6.15) \qquad \frac{D\boldsymbol{u}}{Dt} = \boldsymbol{u}_t + (\boldsymbol{u} \cdot \boldsymbol{\nabla})\boldsymbol{u} = -\frac{1}{\rho}\boldsymbol{\nabla} p,$$

called the *Euler equation*. Physically, the left-hand side is the acceleration of fluid particles, and the right-hand side is the force per unit mass arising from pressure gradients.

Finally, there is the clean-up of the energy equation (6.13). Substitute (6.14) into (6.13). We get

$$\frac{D}{Dt}\left(\frac{1}{2}|\boldsymbol{u}|^2 + e\right) = -\frac{1}{\rho}\boldsymbol{\nabla}\cdot(p\boldsymbol{u}).$$

Next, rewrite it as

$$\boldsymbol{u}\cdot\frac{D\boldsymbol{u}}{Dt} + \frac{De}{Dt} = -\frac{1}{\rho}\boldsymbol{u}\cdot\boldsymbol{\nabla}p - \frac{p}{\rho}\boldsymbol{\nabla}\cdot\boldsymbol{u}.$$

The first terms of the left-hand side and right-hand side cancel by the Euler equation (6.15). Substituting for $\boldsymbol{\nabla}\cdot\boldsymbol{u}$ from (6.14), the remaining terms become

$$\frac{De}{Dt} = \frac{p}{\rho^2}\frac{D\rho}{Dt}$$

or finally,

(6.16)
$$\frac{De}{Dt} + p\frac{D}{Dt}\frac{1}{\rho} = 0.$$

Here, $\frac{1}{\rho}$ is volume per unit mass, so (6.16) says that in a small material region (small compared to the space scale of variations in ρ, $\boldsymbol{u}$ and p but large compared to molecules), the change in internal energy is minus the work done against the pressure force. This is the special case of the first law of thermodynamics in the absence of heat transfer.

It follows from (6.16) that at any material point, the pressure is the *same* function $p = p(\rho)$ of the density for all time: Write (6.16) as

$$\left(e_\rho - \frac{p}{\rho^2}\right)\frac{D\rho}{Dt} + e_p\frac{Dp}{Dt} = 0$$

and let $p = p(\rho)$. We get

$$\left(e_\rho - \frac{p}{\rho^2} + e_p p'(\rho)\right)\frac{D\rho}{Dt} = 0,$$

and this identity is satisfied, independent of what ρ is, if $p(\rho)$ satisfies the ODE

(6.17)
$$e_p p'(\rho) + e_\rho - \frac{p}{\rho^2} = 0.$$

There is a one-parameter family of solutions to (6.17). If the density and pressure are initially *uniform* throughout a material volume, then the pressure $p = p(\rho)$ is the *same* function of the density ρ at all points in that material volume. In this case, the mass conservation PDE (6.8) and the Euler equation (6.15) form a closed system for the density and velocity fields:

$$\text{(6.18)} \qquad \rho_t + (\boldsymbol{u} \cdot \boldsymbol{\nabla})\rho + \rho \boldsymbol{\nabla} \cdot \boldsymbol{u} = 0,$$

$$\text{(6.19)} \qquad \boldsymbol{u}_t + (\boldsymbol{u} \cdot \boldsymbol{\nabla})\boldsymbol{u} = -\frac{p'(\rho)}{\rho}\boldsymbol{\nabla}\rho.$$

Problem 6.1 (Tornado season).   We examine steady circular flows in ideal fluid with $\boldsymbol{u} = u(r)\boldsymbol{e}_\vartheta$. Here, r is the radial coordinate and $\boldsymbol{e}_\vartheta$ the tangential unit vector of cylindrical coordinates.

a) Show that $\rho = \rho(r)$ and $p = p(r)$, independent of ϑ, and determine $\rho'(r)$ in terms of $\rho(r)$ and $u(r)$.

b) The *circulation* at radius r is $\Gamma(r) := 2\pi r u(r)$. We will call our circular flow a *vortex* if $\lim_{r\to\infty} \Gamma(r)$ exists. Assuming that ρ is a uniform constant, find the pressure change from $r = 0$ to $r = \infty$ for a vortex which consists of solid body rotation of angular velocity ω in $r < a$ and constant circulation $2\pi a^2 \omega$ in $r > a$. Estimate this pressure change for $\rho \simeq 1.2$ kg/m^3 (sea-level air density at 20°C) and maximum wind velocity $\omega a \simeq 300$ km/h (tornado speed). What is the upward force on a 100 m^2 roof subjected to this pressure difference? How many metric tons is this force equivalent to $(1\,\text{ton} = 10^3\,\text{kg})$?

Solution.

a) First, look at the time-independent mass conservation PDE,

$$(\boldsymbol{u} \cdot \boldsymbol{\nabla})\rho = -(\boldsymbol{\nabla} \cdot \boldsymbol{u})\rho,$$

which takes the explicit form

$$\frac{u}{r}\rho_\vartheta = -(u_r \boldsymbol{e}_r \cdot \boldsymbol{e}_\vartheta + u\boldsymbol{\nabla} \cdot \boldsymbol{e}_\vartheta)\rho$$
$$= -(0 u_r + 0 u)\rho = 0.$$

Hence, $\rho = \rho(r)$ independent of ϑ. The steady Euler equation,

$$(\boldsymbol{u} \cdot \boldsymbol{\nabla})\boldsymbol{u} = -\frac{1}{\rho}\boldsymbol{\nabla}p,$$

takes the explicit form

$$\text{(6.1-1)} \qquad \frac{u}{r}(u\boldsymbol{e}_\vartheta)_\vartheta = -\frac{1}{\rho}\left(p_r \boldsymbol{e}_r + \frac{p_\vartheta}{r}\boldsymbol{e}_\vartheta\right).$$

In the left-hand side, $e_{\vartheta_\vartheta} = -e_r$; so the left-hand side is the usual centripetal acceleration $-\frac{u^2}{r} e_r$ for a circular orbit of tangential velocity u, and it follows from (6.1-1) that

$$p_\vartheta = 0, \quad \frac{1}{\rho} p_r = \frac{u^2}{r}.$$

The first equation implies that $p = p(r)$ independent of ϑ, and then the second, which expresses balance between centripetal acceleration and pressure force per unit mass, gives

(6.1-2) $$p'(r) = \frac{\rho(r)u^2(r)}{r}.$$

b) We have

$$u(r) = \begin{bmatrix} \omega r & \text{in } 0 < r < a, \\[2mm] \frac{\omega a^2}{r} & \text{in } r > a, \end{bmatrix}$$

and then integration of (6.1-2) with $\rho = \text{constant}$ gives

$$[p]_0^\infty = \rho\omega^2 \left\{ \int_0^a r \, dr + a^4 \int_a^\infty \frac{dr}{r^3} \right\} = \rho\omega^2 a^2.$$

The numerical value is

$$[p]_0^\infty = \left(1.2 \text{ kg/m}^3 \right) \left(\frac{3 \times 10^5 \text{ m}}{3.6 \times 10^3 \text{ s}} \right)^2 = .83 \times 10^4 \text{ N/m}^2.$$

This is about 1/10th of sea-level air pressure, 1.03×10^5 N/m^2, but the effect is going to be very strong: The "lifting force" on our 100 m^2 roof is 0.83×10^6 N. This is equivalent to $0.83 \times 10^6/9.8$ kg, or about 83 metric tons.

Problem 6.2 (Ideal gas equation of state).   *Ideal gas* consists of "point particles" of mass m. In a sufficiently small region of space, the distribution of translational velocities v is independent of position. The mean $u := \langle v \rangle$ defines the velocity field of macroscopic fluid mechanics. In the rest frame of a material point of the macroscopic flow, we have $\langle v \rangle = 0$, and the distribution of e-velocities $v = v \cdot e$ is the same even function $g(v)$ for any unit vector e. The internal energy e per unit mass is just the average kinetic energy per unit mass seen in the local rest frame.

a) Show that $e = \frac{3}{2} \int_{-\infty}^\infty g(v)v^2 \, dv$.

b) We have ideal gas at rest with $\langle v \rangle = 0$ in $x_1 < 0$, and gas particles rebound elastically off the wall at $x_1 = 0$. The pressure p on the wall is the average impulse delivered per unit time per unit area. Show that

$$\frac{p}{\rho} = \int_{-\infty}^{\infty} g(v)v^2 \, dv,$$

so that combining with the result from Part a gives the ideal gas equation of state

(6.2-1)
$$e = \frac{3}{2} \frac{p}{\rho}.$$

c)   Here is a rule of thumb in classical statistical mechanics: "In an equilibrium with absolute temperature T, there is internal energy $\frac{1}{2} k_B T$ for every degree of freedom." The coefficient k_B which converts temperature (in some conventional units such as degrees Kelvin) into energy is called the *Boltzmann constant*. Reformulate the equation of state (6.2-1) as a relation between specific volume $v := 1/$(number density of gas particles), pressure, and absolute temperature.

Solution.

a)   The kinetic energy of one particle is $\frac{m}{2}(v_1^2 + v_2^2 + v_3^2)$. The average values of v_1^2, v_2^2, v_3^2 have common value $\int_{-\infty}^{\infty} g(v)v^2 \, dv$. Hence, the average kinetic energy of a single particle is $\frac{3}{2} m \int_{-\infty}^{\infty} g(v)v^2 \, dv$, and the average kinetic energy per unit mass is

$$e = \frac{3}{2} \int_{-\infty}^{\infty} g(v)v^2 \, dv.$$

b)   The *number* density of particles with e_1-velocities between $v_1 > 0$ and $v_1 + dv_1$ is $\frac{\rho}{m} g(v_1) \, dv_1$, and the e_1-flux of these particles is $\frac{\rho}{m} g(v_1) v_1 \, dv_1$. Each of these particles delivers an impulse $2mv_1$ when it bounces off the wall; so the rate at which the particles deliver impulse per unit time per unit area is the product

$$(2mv_1) \left(\frac{\rho}{m} g(v_1) v_1 \, dv_1 \right) = 2\rho g(v_1) v_1^2 \, dv_1.$$

The contribution from all particles with $v_1 > 0$ (the ones that collide) gives the pressure, so

$$p = 2\rho \int_0^{\infty} g(v_1) v_1^2 \, dv_1 = \rho \int_{-\infty}^{\infty} g(v)v^2 \, dv.$$

c)   A "point" particle in $\mathbb{R}^3$ has three degrees of freedom, so we have $me = \frac{3}{2} k_B T$. Combining with (6.2-1) gives

(6.2-2)
$$\frac{p}{\rho} = \frac{k_B T}{m},$$

and using $v = \frac{m}{\rho}$, we get

(6.2-3) $$pv = k_B T.$$

Problem 6.3 (Bernoulli's theorem and flying over Dagoba).

a) We have *steady* flow of ideal fluid, with ρ and u being time-independent. The reduced equation of state $p = p(\rho)$ applies. *Streamlines* are trajectories of material points, or integral curves of the steady velocity field u. Show that there is a function $h = h(\rho)$ such that $\frac{1}{2}|u|^2 + h(\rho)$ is constant along streamlines. This result is called *Bernoulli's theorem*.

b) For ideal gas, show that $\frac{1}{2}|u|^2 + \frac{5}{2}\frac{k_B T}{m}$ is constant along streamlines. Here, m is the monomer mass of the ideal gas. Suppose we have an airplane flying at uniform velocity in nasty humid air, completely saturated with water vapor. What do we see happening just above the upper wing surfaces?

Solution.

a) With the help of the Euler equation, we compute

(6.3-1) $$\frac{D}{Dt}\left(\frac{1}{2}|u|^2\right) = u \cdot \frac{Du}{Dt} = -\frac{u \cdot \nabla p}{\rho} = -u \cdot \left(\frac{p'(\rho)}{\rho}\nabla\rho\right).$$

Now define $h(\rho)$ as an antiderivative of $\frac{p'(\rho)}{\rho}$, so that

(6.3-2) $$h'(\rho) = \frac{p'(\rho)}{\rho},$$

and then (6.3-1) becomes

$$\frac{D}{Dt}\left(\frac{1}{2}|u|^2\right) = -(u \cdot \nabla)h(\rho).$$

In *steady* flow, $(u \cdot \nabla)h(\rho) = \frac{Dh}{Dt}$ and so

$$\frac{D}{Dt}\left(\frac{1}{2}|u|^2 + h(\rho)\right) = 0.$$

Hence, $\frac{1}{2}|u|^2 + h(\rho)$ is constant along particle paths, which for steady flow are the streamlines.

b) Given the ideal gas equation of state (6.2-1), the energy identity (6.16) takes the explicit form

$$\frac{3}{2}\frac{D}{Dt}\left(\frac{p}{\rho}\right) + p\frac{D}{Dt}\left(\frac{1}{\rho}\right) = 0$$

or, upon rearranging,

$$\frac{1}{p}\frac{Dp}{Dt} = \frac{5}{3}\frac{1}{\rho}\frac{D\rho}{Dt}.$$

Hence, along particle paths,

(6.3-3)
$$\frac{p}{\rho^{\frac{5}{3}}} = C,$$

where C is a constant independent of t. Here, we are considering the case where C is the *same* constant for all material points, so we can write

$$p = p(\rho) = (C\rho)^{\frac{5}{3}};$$

then (6.3-2) reads

$$h'(\rho) = \frac{5}{3}C^{\frac{5}{3}}\rho^{-\frac{1}{3}},$$

and integration gives

(6.3-4)
$$h(\rho) = \frac{5}{2}C^{\frac{5}{3}}\rho^{\frac{2}{3}} = \frac{5}{2}\frac{p}{\rho}$$

modulo an additive constant. The second equality comes from elimination of C by the relation $p = (C\rho)^{5/3}$. Given (6.3-4), it follows from Part a that $\frac{1}{2}|\boldsymbol{u}|^2 + \frac{5}{2}\frac{p}{\rho}$ is constant along streamlines. Finally, $\frac{p}{\rho} = \frac{k_B T}{m}$ as in (6.2-2), so $\frac{1}{2}|\boldsymbol{u}|^2 + \frac{5}{2}\frac{k_B T}{m}$ is constant along streamlines.

The streamlines of steady flow about our wing in its rest frame are depicted in Figure 6.2.

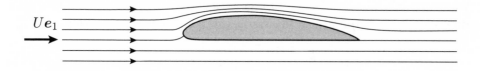

Figure 6.2.

There is uniform flow $\boldsymbol{u} \to U\boldsymbol{e}_1$ as $x_1 \to -\infty$. The flow speed above the upper curved wing surface is *faster* than U, so the temperature T drops *below* the ambient value at $x_1 \to \infty$, and "fog" condenses out of this oversaturated air. The same physics explains why filling station air hoses often "spit" a nasty mixture of air and water.

Problem 6.4 (Circulation and vorticity).

a) Let $C(t)$ be a closed material curve in ideal fluid with the simplified equation of state, $p = p(\rho)$. Show that the *circulation*

$$(6.4\text{-}1) \qquad\qquad \Gamma := \int_{C(t)} \boldsymbol{u} \cdot d\boldsymbol{x}$$

is independent of time t.

b) Show that the *vorticity field* $\boldsymbol{\omega} := \nabla \times \boldsymbol{u}$ of the velocity field $\boldsymbol{u}$ satisfies the PDE

$$(6.4\text{-}2) \qquad\qquad \boldsymbol{\omega}_t + (\boldsymbol{u} \cdot \nabla)\boldsymbol{\omega} = (\boldsymbol{\omega} \cdot \nabla)\boldsymbol{u} - (\nabla \cdot \boldsymbol{u})\boldsymbol{\omega}.$$

(Hint: Review Problems 1.13 and 1.14 on convected vectors and the vector analogs of convected densities.)

Solution.

a) Let $\boldsymbol{x}(s,t)$ be a parametric representation of the material loop $C(t)$, with s in $[0,1]$ labeling material points; the closure of the loop is represented by $\boldsymbol{x}(0,t) = \boldsymbol{x}(1,t)$. The parametric form of the line integral (6.4-1) is

$$(6.4\text{-}3) \qquad\qquad \Gamma = \int_0^1 \boldsymbol{u}(\boldsymbol{x}(s,t),t) \cdot \boldsymbol{\tau}(s,t)\, ds,$$

where $\boldsymbol{\tau}(s,t) := \boldsymbol{x}_s(s,t)$ is the tangent vector of $C(t)$ at $\boldsymbol{x}(x,t)$. We time differentiate (6.4-3) and use the result of Problem 1.13a to obtain

$$(6.4\text{-}4)
\begin{aligned}
\dot{\Gamma} &= \int_0^1 \left\{ \frac{D\boldsymbol{u}}{Dt} \cdot \boldsymbol{\tau} + \boldsymbol{u} \cdot (\boldsymbol{\tau} \cdot \nabla)\boldsymbol{u} \right\} ds \\
&= \int_0^1 \boldsymbol{\tau} \cdot \left\{ \frac{D\boldsymbol{u}}{Dt} + \nabla \left(\frac{1}{2}|\boldsymbol{u}|^2 \right) \right\} ds.
\end{aligned}$$

By the Euler equation (6.15) with $p = p(\rho)$,

$$\frac{D\boldsymbol{u}}{Dt} = -\frac{1}{\rho}\nabla p = -\frac{p'(\rho)}{\rho}\nabla\rho = -\nabla h(\rho),$$

where $h(\rho)$ is an antiderivative of $\frac{p(\rho)}{\rho}$, as in Problem 6.3. Hence, (6.4-4) becomes

$$\begin{aligned}
\dot{\Gamma} &= \int_0^1 \boldsymbol{\tau} \cdot \nabla \left(-h(\rho) + \frac{1}{2}|\boldsymbol{u}|^2 \right) ds \\
&= \int_{C(t)} \nabla \left(-h(\rho) + \frac{1}{2}|\boldsymbol{u}|^2 \right) \cdot d\boldsymbol{x} = 0.
\end{aligned}$$

b) Let $S(t)$ be any two-dimensional material surface whose boundary is a closed curve $C(t)$. By Stokes' theorem,

$$\int_{S(t)} \boldsymbol{\omega} \cdot \boldsymbol{n} \, da = \int_{C(t)} \boldsymbol{u} \cdot d\boldsymbol{x} = \Gamma,$$

the time-independent circulation around $\partial S = C$. Hence, for all such material regions $S(t)$, the surface integral of $\boldsymbol{\omega}$ over $S(t)$ is independent of time. Now recall Problem 1.14 on the two vector analogs of convected densities in $\mathbb{R}^3$. The vorticity field is a realization of the second (known by geometers as a "convected two-form"). In Problem 1.14b we showed that such a $\boldsymbol{\omega}$ satisfies the vorticity PDE (6.4-2).

We review a conventional derivation of the vorticity PDE, based on (rather mechanical) use of vector calculus identities. First, write the Euler PDE for $p = p(\rho)$ as we did in Part a,

$$\boldsymbol{u}_t + (\boldsymbol{u} \cdot \boldsymbol{\nabla})\boldsymbol{u} = -\boldsymbol{\nabla}h(\rho),$$

and then take the curl. Since $\boldsymbol{\nabla} \times \boldsymbol{\nabla}h(\rho)$ is automatically zero, we have

(6.4-5) $$\boldsymbol{\omega}_t + \boldsymbol{\nabla} \times ((\boldsymbol{u} \cdot \boldsymbol{\nabla})\boldsymbol{u}) = 0.$$

In the identity

(6.4-6) $$\boldsymbol{\nabla}(\boldsymbol{a} \cdot \boldsymbol{b}) = (\boldsymbol{a} \cdot \boldsymbol{\nabla})\boldsymbol{b} + (\boldsymbol{b} \cdot \boldsymbol{\nabla})\boldsymbol{a} + \boldsymbol{a} \times (\boldsymbol{\nabla} \times \boldsymbol{b}) + \boldsymbol{b} \times (\boldsymbol{\nabla} \times \boldsymbol{a}),$$

we take $\boldsymbol{a} = \boldsymbol{b} = \boldsymbol{u}$ to obtain

$$\boldsymbol{\nabla}\left(\frac{1}{2}\boldsymbol{u} \cdot \boldsymbol{u}\right) = (\boldsymbol{u} \cdot \boldsymbol{\nabla})\boldsymbol{u} + \boldsymbol{u} \times \boldsymbol{\omega},$$

and taking the curl then gives

(6.4-7) $$\boldsymbol{\nabla} \times ((\boldsymbol{u} \cdot \boldsymbol{\nabla})\boldsymbol{u}) = -\boldsymbol{\nabla} \times (\boldsymbol{u} \times \boldsymbol{\omega}).$$

We compute the right-hand side of (6.4-7) by taking $\boldsymbol{a} = \boldsymbol{u}$ and $\boldsymbol{b} = \boldsymbol{\omega}$ in a second identity

(6.4-8) $$\boldsymbol{\nabla} \times (\boldsymbol{a} \times \boldsymbol{b}) = (\boldsymbol{b} \cdot \boldsymbol{\nabla})\boldsymbol{a} - (\boldsymbol{a} \cdot \boldsymbol{\nabla})\boldsymbol{b} + \boldsymbol{a}\boldsymbol{\nabla} \cdot \boldsymbol{b} - \boldsymbol{b}\boldsymbol{\nabla} \cdot \boldsymbol{a};$$

then (6.4-7) becomes

$$\boldsymbol{\nabla} \times ((\boldsymbol{u} \cdot \boldsymbol{\nabla})\boldsymbol{u}) = (\boldsymbol{u} \cdot \boldsymbol{\nabla})\boldsymbol{\omega} - (\boldsymbol{\omega} \cdot \boldsymbol{\nabla})\boldsymbol{u} + (\boldsymbol{\nabla} \cdot \boldsymbol{u})\boldsymbol{\omega}.$$

Finally, (6.4-5) becomes

$$\boldsymbol{\omega}_t + (\boldsymbol{u} \cdot \boldsymbol{\nabla})\boldsymbol{\omega} = (\boldsymbol{\omega} \cdot \boldsymbol{\nabla})\boldsymbol{u} - (\boldsymbol{\nabla} \cdot \boldsymbol{u})\boldsymbol{\omega}.$$

Problem 6.5 (Heating due to shock wave).   There are weak solutions of the ideal fluid equations which have ρ, p, and the normal component of $\boldsymbol{u}$ discontinuous across two-dimensional surfaces in $\mathbb{R}^3$. We will look at the simplest example of such a *shock wave*. The shock is the surface $x_1 = 0$. (We are in the shock's rest frame.) The state variables of the flow in $x_1 < 0$ are $\rho \equiv \rho_1$, $p \equiv p_1$, and $\boldsymbol{u} \equiv u_1 \boldsymbol{e}_1$, where ρ_1, p_1 and u_1 are positive constants. In $x_1 > 0$, we have another set of constants: $\rho \equiv \rho_2$, $p \equiv p_2$ and $\boldsymbol{u} \equiv u_2 \boldsymbol{e}_1$.

a) Write down jump conditions which guarantee that the shock is not a source of mass, momentum, or energy.

b) For simplicity, we will henceforth use the ideal gas equation of state, $e = \frac{3}{2}\frac{p}{\rho}$, which simplifies the energy jump condition. Show that $r := \frac{\rho_2}{\rho_1}$ and $s := \frac{p_2}{p_1}$ satisfy

$$(6.5\text{-}1) \qquad\qquad\qquad r = \frac{1 + 4s}{4 + s}.$$

c) Recall that in a smooth flow of ideal gas, $\frac{p^{3/5}}{\rho}$ is constant at a material point, as we showed in Problem 6.3. Show that a material point passing through our shock wave *cannot* preserve its value of $\frac{p^{3/5}}{\rho}$. (Hint: (6.5-1) is *one* relation between r and s. A proposed invariance of $\frac{p^{3/5}}{\rho}$ would give another.)

d) Here is the standard characterization of the nonideal energetics that happens inside the (unresolved) shock layer: After a material region crosses the shock layer, its temperature is always *greater* than the value that is consistent with no heat transfer to it. Let us call this added increment of temperature δT. Show that $\delta T > 0$ implies $p_2 > p_1$ and $\rho_2 > \rho_1$. Compute $\frac{\delta T}{T_1}$ as a function of the shock strength $\sigma := \frac{p_2}{p_1} - 1 = s - 1$. Show that $\frac{\delta T}{T_1}$ scales like σ^3 in the limit $\sigma \to 0^+$.

e) Show that the pre-shock velocity u_1 is greater than $c_1 := \sqrt{\frac{5}{3}\frac{p_1}{\rho_1}}$.

Solution.

a) The jump conditions represent continuity of the relevant fluxes:

$$(6.5\text{-}2) \qquad \begin{array}{ll} \text{Mass} & [\![\rho u]\!] = \rho_2 u_2 - \rho_1 u_1 = 0, \\ \text{Momentum} & [\![\rho u^2 + p]\!] = 0, \\ \text{Energy} & [\![(\frac{1}{2}\rho u^2 + \rho e + p)u]\!] = 0. \end{array}$$

b) Given the ideal gas equation of state, the energy balance condition becomes

$$(6.5\text{-}3) \qquad\qquad\qquad \left[\!\!\left[\left(\frac{1}{2}\rho u^2 + \frac{5}{2}p\right) u \right]\!\!\right] = 0.$$

Let f be the common value of the mass fluxes $\rho_1 u_1$ and $\rho_2 u_2$, and set $u_1 = \frac{f}{\rho_1}$ and $u_2 = \frac{f}{\rho_2}$ in the momentum balance condition (second equation of (6.5-2)) and energy balance condition (6.5-3). We get

$$\left[\frac{f^2}{\rho} + p\right] = 0, \qquad \left[\frac{f^2}{\rho^2} + 5\frac{p}{\rho}\right] = 0,$$

and these are simplified to

(6.5-4) $$f^2 = \rho_1\rho_2 \frac{p_2 - p_1}{\rho_2 - \rho_1} = 5\rho_1\rho_2 \frac{\rho_1 p_2 - \rho_2 p_1}{\rho_2^2 - \rho_1^2}.$$

The identity based on the second equality is

$$(\rho_1 + \rho_2)(p_2 - p_1) = 5(\rho_1 p_2 - \rho_2 p_1)$$

or, in terms of $r := \frac{\rho_2}{\rho_1}$ and $s := \frac{p_2}{p_1}$,

$$(1 + r)(s - 1) = 5(s - r);$$

so finally, $r = \frac{1+4s}{4+s}$ as in (6.5-1).

c) If we try to impose $\frac{p_2^{3/5}}{\rho_2} = \frac{p_1^{3/5}}{\rho_1}$, we get $r = s^{3/5}$. Figure 6.3 shows the graphs of the two relations $r = \frac{1+4s}{4+s}$ and $r = s^{3/5}$.

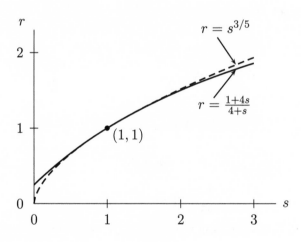

Figure 6.3.

The graphs intersect *only* at $(r, s) = (1, 1)$. Since $r = \frac{1+4s}{4+s}$ is the *physical* relation between r and s, $r = s^{\frac{3}{5}}$ cannot happen for $r \neq 1$. But $r = 1$ means no shock.

d) We have $e = \frac{3}{2}\frac{p}{\rho} \propto T$, so

(6.5-5) $$\frac{T_2}{T_1} = \frac{s}{r} = \frac{(4+s)s}{1+4s}.$$

If there is no heat transfer, so that the value of $\frac{p^{3/5}}{\rho}$ is preserved across the shock and $r = s^{3/5}$, then the post-shock temperature would be T_2' determined by

(6.5-6)
$$\frac{T_2'}{T_1} = \frac{s}{s^{\frac{3}{5}}} = s^{\frac{2}{5}}.$$

It follows from (6.5-5) and (6.5-6) that

$$\frac{\delta T}{T_1} = \frac{T_2}{T_1} - \frac{T_2'}{T_1} = \frac{(4+s)s}{1+4s} - s^{\frac{2}{5}}$$

or, using the shock strength $\sigma = s - 1$,

$$\frac{\delta T}{T_1} = \frac{(5+\sigma)(1+\sigma)}{5+4\sigma} - (1+\sigma)^{\frac{2}{5}}.$$

Figure 6.4 shows the graph in $0 < \sigma < 2$.

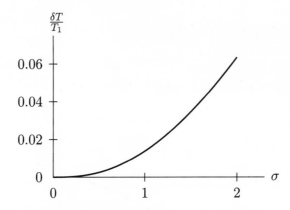

Figure 6.4.

There is *cubic* contact with zero at $\sigma = 0$:

$$\frac{\delta T}{T_1} = \frac{4}{125}\sigma^3 + O(\sigma^4) \quad \text{as } \sigma \to 0^+.$$

For $-1 < \sigma < 0$, corresponding to $0 < p_2 < p_1$, we would have $\delta T < 0$, which is nonphysical. Hence $p_2 > p_1$, and by the monotone increasing character of the relation (6.5-4) between r and s, $\rho_2 > \rho_1$.

e) The first equality in (6.5-4) can be re-expressed as

$$u_1^2 = \frac{f^2}{\rho_1^2} = \frac{\rho_2}{\rho_1}\frac{p_2 - p_1}{\rho_2 - \rho_1} = \frac{p_1}{\rho_1}\frac{r(s-1)}{r-1},$$

and upon setting $r = \frac{1+4s}{4+s}$ as in (6.5-1) and $\frac{p_1}{\rho_1} = \frac{3}{5}c_1^2$, we get

$$u_1^2 = c_1^2\frac{1+4s}{5}.$$

Introducing the shock strength $\sigma := s - 1 > 0$, we have

$$u_1^2 = c_1^2 \left(1 + \frac{4}{5}\sigma\right) > c_1^2.$$

We will soon realize the significance of c_1 as the pre-shock sound speed. In the rest frame of the pre-shock gas, the shock moves with speed $c_1\sqrt{1 + \frac{4}{5}\sigma} > c_1$, so shocks are "supersonic".

Stable rest state of ideal fluid. We examine small perturbations from a homogeneous fluid at rest, with $\rho \equiv \rho_0$, a positive constant, and $\boldsymbol{u} \equiv 0$. The linearizations of (6.18) and (6.19) for the density perturbation $\eta := \rho - \rho_0$ and $\boldsymbol{u}$ are

$$\eta_t + \rho_0 \boldsymbol{\nabla} \cdot \boldsymbol{u} = 0, \qquad \boldsymbol{u}_t + \frac{p'(\rho_0)}{\rho_0}\boldsymbol{\nabla}\eta = 0.$$

Take the time derivative of the first equation and the divergence of the second, and then combine these to get

(6.20) $$\eta_{tt} - p'(\rho_0)\Delta\eta = 0.$$

If $p'(\rho_0) < 0$, small density perturbations rapidly amplify in time, and the homogeneous rest state of fluid is unstable. To see this, define $\sigma > 0$ such that $\sigma^2 = -p'(\rho_0)$; σ has the units of velocity, and (6.20) can be nondimensionalized so that the unit of time is the unit of length divided by σ. The dimensionless version of (6.20) is Laplace's equation

(6.21) $$\eta_{tt} + \Delta\eta = 0$$

in $\mathbb{R}^3 \times \mathbb{R} = \mathbb{R}^4$. Let r be the distance from some origin of spatial $\mathbb{R}^3$ and let $s = \sqrt{t^2 + r^2}$ be the distance from the origin ($r = 0$, $t = 0$) in $\mathbb{R}^4$. Radial solutions of (6.21) with $\eta = \eta(s)$ satisfy

$$\eta_{ss} + \frac{3}{s}\eta_s = 0,$$

and there is a solution

(6.22) $$\frac{1}{2s^2} = \frac{1}{2}\frac{1}{t^2 + r^2}$$

which vanishes as $s \to \infty$. This solution looks like a mass perturbation concentrated about $r = 0$, but in fact the total mass diverges:

$$\int_0^\infty \frac{1}{2}\frac{1}{t^2 + r^2}4\pi r^2 \, dr = +\infty.$$

However, the t-derivative of (6.22),

(6.23) $$\eta := \frac{-t}{t^2 + r^2} = -\frac{1}{t^3} \frac{1}{(1 + r^2/t^2)^2},$$

is also a solution of (6.21). For any $t < 0$, (6.23) looks like a mass concentration about $r = 0$, and now the total mass perturbation is finite and *constant* in t:

$$\int_0^\infty \eta 4\pi r^2 \, dr = 4\pi \int_0^\infty \frac{-tr^2}{(t^2 + r^2)^2} \, dr$$

$$= 4\pi \int_0^\infty \frac{s^2}{(1 + s^2)^2} \, ds = \text{ positive constant.}$$

For $t < 0$, (6.23) represents a density perturbation whose total mass collapses to the origin as $t \to 0^-$. The mechanical process is clear: If $p'(\rho_0) < 0$, the pressure is smaller inside the concentration, where the density is higher. Hence, the pressure force further concentrates the mass, which leads to yet lower pressure inside and so on—*implosion*. In summary, fluids whose homogeneous rest configurations are stable cannot have $p'(\rho) < 0$.

Finite propagation speed and wavefronts. Henceforth, assume $c^2 := p'(\rho_0) > 0$. Then (6.20) is a wave equation

$$\eta_{tt} - c^2 \Delta \eta = 0$$

with propagation speed c. The intuitive statement that "information propagates at speed c" means that the domain of influence associated with any event $(\boldsymbol{X}, T)$ in spacetime is the cone defined by

$$t - T = \frac{|\boldsymbol{x} - \boldsymbol{X}|}{c}.$$

This cone is depicted in Figure 6.5a.

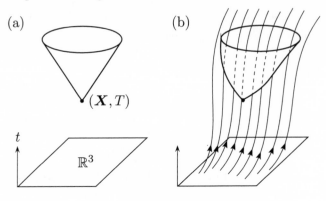

Figure 6.5.

The finite propagation speed of information is also a property of the full nonlinear PDEs (6.18) and (6.19). In particular, consider *wavefronts*. A wavefront is a time sequence $S(t)$ of two-dimensional surfaces in spatial $\mathbb{R}^3$ such that ρ and $\boldsymbol{u}$ are continuous across $S(t)$ but their normal derivatives have jump discontinuities. Let $\boldsymbol{n}$ denote the unit normal on $S(t)$, and v the normal velocity. Denote the jump in the normal derivative of a function f by $[f_n]$. It follows from (6.18) and (6.19) that $[\rho_n]$ and $[\boldsymbol{u}_n]$ satisfy

$$(6.24) \qquad (-v + \boldsymbol{u} \cdot \boldsymbol{n})[\rho_n] + \rho \boldsymbol{n} \cdot [\boldsymbol{u}_n] = 0,$$

$$(6.25) \qquad (-v + \boldsymbol{u} \cdot \boldsymbol{n})[\boldsymbol{u}_n] + \frac{p'(\rho)}{\rho}[\rho_n]\boldsymbol{n} = 0.$$

Here are the calculations behind (6.24) and (6.25). Let $\boldsymbol{x}(t)$ be a point on the wavefront. Evaluate the mass conservation PDE (6.18) at $\boldsymbol{x} = \boldsymbol{x}(t) + \varepsilon \boldsymbol{n}$ and at $\boldsymbol{x} = \boldsymbol{x}(t) - \varepsilon \boldsymbol{n}$, where $\boldsymbol{n}$ is the unit normal of $S(t)$ at $\boldsymbol{x}(t)$. Subtract these two equations and take the limit $\varepsilon \to 0$. We get

$$(6.26) \qquad [\rho_t] + \boldsymbol{u} \cdot [\boldsymbol{\nabla}\rho] + \rho[\boldsymbol{\nabla} \cdot \boldsymbol{u}] = 0,$$

where we have used the continuity of $\boldsymbol{u}$ and ρ across $S(t)$. Since tangential derivatives are continuous across $S(t)$, $[\boldsymbol{\nabla}\rho] = [\rho_n]\boldsymbol{n}$ and hence

$$(6.27) \qquad \boldsymbol{u} \cdot [\boldsymbol{\nabla}\rho] = (\boldsymbol{u} \cdot \boldsymbol{n})[\rho_n].$$

This takes care of the second term in (6.26). To deal with $[\boldsymbol{\nabla} \cdot \boldsymbol{u}]$, employ cartesian coordinates with the origin at $\boldsymbol{x}(t)$, $\boldsymbol{e}_1 = \boldsymbol{n}$, and $\boldsymbol{e}_2$, $\boldsymbol{e}_3$ in the tangent plane of $S(t)$ at $\boldsymbol{x}(t)$. We have $\boldsymbol{\nabla} \cdot \boldsymbol{u} = \partial_1 u_1 + \partial_2 u_2 + \partial_3 u_3$, and $\partial_2 u_2$ and $\partial_3 u_3$ are continuous across $S(t)$, so $[\boldsymbol{\nabla} \cdot \boldsymbol{u}] = [\partial_1 u_1] = [\partial_1(\boldsymbol{u} \cdot \boldsymbol{e}_1)] = [\partial_1 \boldsymbol{u}] \cdot \boldsymbol{e}_1 = [\partial_n \boldsymbol{u}] \cdot \boldsymbol{n}$. Hence,

$$(6.28) \qquad \rho[\boldsymbol{\nabla} \cdot \boldsymbol{u}] = \rho \boldsymbol{n} \cdot [\boldsymbol{u}_n],$$

and this accounts for the third term in (6.26). This leaves only $[\rho_t]$ in (6.26). By the chain rule,

$$\frac{d}{dt}\{\rho(\boldsymbol{x}(t) + \varepsilon\boldsymbol{n}, t) - \rho(\boldsymbol{x}(t) - \varepsilon\boldsymbol{n}, t)\}$$
$$= (\rho_t + (\dot{\boldsymbol{x}} + \varepsilon\boldsymbol{n}) \cdot \boldsymbol{\nabla}\rho)(\boldsymbol{x}(t) + \varepsilon\boldsymbol{n}, t) - (\rho_t + (\dot{\boldsymbol{x}} - \varepsilon\boldsymbol{n}) \cdot \boldsymbol{\nabla}\rho)(\boldsymbol{x}(t) - \varepsilon\boldsymbol{n}, t).$$

In the limit $\varepsilon \to 0$, the left-hand side vanishes by the continuity of ρ across $S(t)$. Hence,

$$[\rho_t] + \dot{\boldsymbol{x}} \cdot [\boldsymbol{\nabla}\rho] = 0$$

or, using $[\boldsymbol{\nabla}\rho] = [\rho_n]\boldsymbol{n}$,

$$(6.29) \qquad [\rho_t] = -\dot{\boldsymbol{x}} \cdot \boldsymbol{n}[\rho_n] = -v[\rho_n],$$

where $v := \dot{\boldsymbol{x}} \cdot \boldsymbol{n}$ is the normal velocity of $S(t)$ at $\boldsymbol{x}(t)$. Substituting (6.27)–(6.29) into (6.26) gives the identity

$$-v[\rho_n] + (\boldsymbol{u} \cdot \boldsymbol{n})[\rho_n] + \rho \boldsymbol{n} \cdot [\boldsymbol{u}_n] = 0,$$

which simplifies to (6.24). The derivation of (6.25) is similar: Evaluate the Euler equation (6.19) at $x = x(t) + \varepsilon n$ and $x = x - \varepsilon n$; then take the $\varepsilon \to 0$ limit of the difference.

Equations (6.24) and (6.25) form a system of linear, homogeneous equations for $[\rho_n]$ and $[u_n]$. There is a solvability condition: Take the inner product of (6.25) with n to get

$$(6.30) \qquad (-v + u \cdot n) n \cdot [u_n] + \frac{p'(\rho)}{\rho} [\rho_n] = 0.$$

The system (6.24)–(6.30) has nontrivial solutions for $[\rho_n]$ and $n \cdot [u_n]$ if and only if

$$(6.31) \qquad (v - u \cdot n)^2 = p'(\rho).$$

The quantity $v - u \cdot n$ is the normal velocity of the wavefront as seen by an observer moving with the fluid at velocity u. (6.31) says that this "relative" normal velocity is $\pm\sqrt{p'(\rho)}$. This is just what one expects, based on the propagation speed $\sqrt{p'(\rho)}$ in fluid at rest, plus Galilean invariance. The determination of the normal velocity v according to (6.31) is in essence the evolution equation of the wavefront. Wavefronts emanating from a single point in spacetime have the same topology as the cone in Figure 6.5a, but in general they are distorted by the nonuniformity of ρ and u, as shown in Figure 6.5b.

Problem 6.6 (Sonic boom).

a)   We consider steady flows of ideal fluid with the reduced equation of state, $p = p(\rho)$, being an increasing function of ρ. Show that stationary (non-moving) wavefronts can live only in regions of supersonic flow with $|u| > c(\rho) := \sqrt{p'(\rho)}$. In the case of two-dimensional flow, what are the possible directions of a wavefront's tangent vector, in relation to the basal (supersonic) velocity u? What is the angle α of the Mach cone induced by a disturbance at $x = 0$ to uniform supersonic flow $u e_1$, as depicted in Figure 6.6?

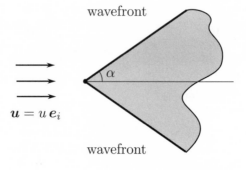

wavefront

$u = u\, e_i$

wavefront

Figure 6.6.

b) From the equations of time-independent flow, derive linearized equations for the density perturbation $\eta := \rho - \rho_0$ and the velocity perturbation $\boldsymbol{v} := \boldsymbol{u} - u_0\boldsymbol{e}_1$ from uniform flow with density ρ_0 and velocity $u_0\boldsymbol{e}_1$, where ρ_0 and u_0 are positive constants. Derive a single second-order PDE for the density perturbation η.

c) We analyze the perturbation to uniform flow induced by a symmetric wing. The wing extends indefinitely in the x_3 direction, so we are looking at a two-dimensional flow in the (x_1, x_2) plane, as depicted in Figure 6.7.

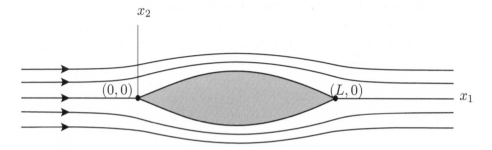

Figure 6.7.

By symmetry about the x_1-axis, it is sufficient to formulate a boundary value problem in $x_2 > h(x_1)$, where $h(x_1) = 0$ outside of $(0, L)$, and $h(x_1) > 0$ in $(0, L)$ represents the upper wing surface. Determine effective values of $v_2(x_1, 0)$ in the thin-wing limit, max $h \ll L$. (Hint: In $0 < x_1 < L$, the physical boundary conditions happen on $x_2 = h(x_1)$, and we obtain effective boundary conditions on $x_2 = 0$ by analytic continuation.) Determine effective values of $\partial_2 \eta(x_1, 0)$.

d) We are going to analyze the wing-induced perturbation of *supersonic* flow $\boldsymbol{u} = u_0\boldsymbol{e}_1$, $u_0 > c(\rho_0)$. First, determine the *physically relevant* density perturbation which satisfies the PDE of Part b in $x_2 > 0$, and has the effective values of $\partial_2 \eta$ derived in Part c. (Hint: There are not enough boundary conditions to select a unique solution! You need to impose some additional constraint on the solution. Think about domains of dependence and the Mach cone in Part a.) Next, determine v_2 in $x_2 > 0$, and asymptotically construct the streamlines of the flow in the thin-wing limit. Sketch these streamlines in the (x_1, x_2) plane. In particular, the region of flow influenced by the wing should be clear from your sketch.

Suppose you are a material particle coming from $x_1 = -\infty$ at fixed $x_2 > 0$. When you enter the wing-induced "wave" (the support of η and $\boldsymbol{v}$), are you crossing a "wavefront" where density and velocity are continuous but normal derivatives discontinuous, or is it a *shock* with η and $\boldsymbol{v}$ nonzero after you cross? That depends. On what? Discuss.

Solution.

a) For stationary wavefronts with normal velocity $v = 0$, the "wavefront condition" (6.31) reduces to

$$(6.6\text{-}1) \qquad\qquad \boldsymbol{u} \cdot \boldsymbol{n} = c(\rho).$$

A possible minus sign on the right-hand side is absorbed by the orientation of $\boldsymbol{n}$. (6.6-1) has solutions for $\boldsymbol{n}$ only if $|\boldsymbol{u}| > c$. To determine possible tangent vectors of a wavefront at a given point $\boldsymbol{x}$ in two-dimensional flow, take $\boldsymbol{e}_1$ in the $\boldsymbol{u}(\boldsymbol{x})$ direction so that $\boldsymbol{u} = u\boldsymbol{e}_1$, where $u = |\boldsymbol{u}(\boldsymbol{x})| > c$. Then (6.6-1) reduces to $un_1 = c$ or $u\tau_2 = c$ where τ_2 is the $\boldsymbol{e}_2$-component of the wavefront tangent vector. Hence, $\tau_2 = \frac{c}{u} = \frac{1}{m}$, where $m := \frac{u}{c} > 1$ is the *Mach number* of the flow at $\boldsymbol{x}$. Finally, $\boldsymbol{\tau} = \frac{1}{m}(\pm\sqrt{m^2 - 1}\,\boldsymbol{e}_1 + \boldsymbol{e}_2)$. The geometric construction of wavefront tangent lines through $\boldsymbol{x}$ is depicted in Figure 6.8. The shaded region is a portion of the Mach cone emanating from a "disturbance" at $\boldsymbol{x}$. The angulation α is the root of $\sin\alpha = \frac{1}{m}$ with $0 < \alpha < \frac{\pi}{2}$.

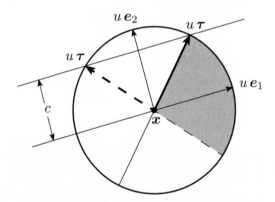

Figure 6.8.

b) The steady flow equations are (6.15) and (6.16) with $\rho_t, \boldsymbol{u}_t = 0$. The linearized equations for $\eta := \rho - \rho_0$ and $\boldsymbol{v} := \boldsymbol{u} - u_0\boldsymbol{e}_1$ are

$$(6.6\text{-}2) \qquad\qquad u_0^2 \partial_1 \eta + \rho_0 u_0 \boldsymbol{\nabla} \cdot \boldsymbol{v} = 0,$$

$$(6.6\text{-}3) \qquad\qquad \rho_0 u_0 \partial_1 \boldsymbol{v} + c^2 \boldsymbol{\nabla}\eta = 0.$$

Here, $c := \sqrt{p'(\rho_0)}$ is the unperturbed speed of sound. Taking the x_1-derivative of (6.6-2) and the divergence of (6.6-3) and subtracting, we get the PDE for η,

$$(6.6\text{-}4) \qquad\qquad m^2 \partial_{11}\eta - \Delta\eta = 0,$$

where $m := \frac{c}{u_0} > 1$ is the Mach number of the unperturbed flow. In the $\mathbb{R}^2$ case (no x_3 dependence), (6.6-4) reduces to

$$(6.6\text{-}5) \qquad \partial_{22}\eta - (m^2 - 1)\partial_{11}\eta = 0.$$

c) The normal component of $\boldsymbol{u} = u_0 \boldsymbol{e}_1 + \boldsymbol{v}$ vanishes on $x_2 = h(x_1)$. The unit normal $\boldsymbol{n}$ at $(x_1, h(x_1))$ is proportional to $\boldsymbol{e}_2 - h'(x_1)\boldsymbol{e}_1$, so the exact boundary condition is

$$(6.6\text{-}6) \qquad v_2(x_1, h(x_1)) - h'(x_1)(u_0 + v_1(x_1, h(x_1))) = 0.$$

A formal way to determine the thin-wing limit, max $h \ll L$, is to multiply $h(x)$ by a positive *gauge parameter* ε. We expect velocity perturbations to scale in proportion to wing thickness, so v_1 and v_2 are also multiplied by ε. The scaled version of (6.6-6) is

$$v_2(x_1, \varepsilon h(x_1)) - h'(x_1)(u_0 + \varepsilon v_1(x_1, \varepsilon h(x_1))) = 0,$$

and the $\varepsilon \to 0$ limit is

$$(6.6\text{-}7) \qquad v_2(x_1, 0) = u_0 h'(x_1).$$

To find effective values of $\partial_1 \eta(x_1, 0)$, look at the $\boldsymbol{e}_2$-component of the linearized Euler equation (6.6-3),

$$(6.6\text{-}8) \qquad \rho_0 u_0 \partial_1 v_2 + c^2 \partial_2 \eta = 0.$$

Evaluate on $x_2 = 0$ and substitute $\partial_1 v_2 = u_0 h''(x_1)$ from (6.6-7) to get

$$(6.6\text{-}9) \qquad \partial_2 \eta = -\rho_0 m^2 h''(x_1).$$

d) For $m > 1$, (6.6-5) is a wave equation. Since boundary values on $\partial_2 \eta$ are given along the x_1-axis, we think of x_1 as "space" and x_2 as "time". In this context, the boundary condition (6.6-9) looks like an "initial condition".

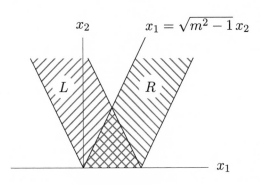

Figure 6.9.

But by itself it is not enough; in the standard initial value problem for the wave equation, we pose given values of $\eta(x_1, 0)$ as well. To select the physical solution, we look at the domain of dependence in $x_2 > 0$ upon the given (generally nonzero) values of $\partial_2 \eta(x_1, 0)$ in $0 < x_1 < L$. In Figure 6.9, the regions of support of right- and left-traveling waves consistent with the initial data are denoted by R and L.

Physically, the support of the perturbations is expected to be *down-stream* from the leading edge $(0,0)$ (like the Mach cone in Part a). Hence, the left-traveling wave, which goes upstream, isn't going to happen. The physical *wake* (the support of η and $\boldsymbol{v}$) is the region R of the right-traveling wave. The *right-traveling wave* solution for η consistent with the initial condition (6.6-9) is

$$(6.6\text{-}10) \qquad \eta = \frac{\rho_0 m^2}{\sqrt{m^2-1}}\{h'(x_1 - \sqrt{m^2-1}\,x_2) + C\}$$

in R, and zero outside of R. Here, C is a constant. We claim that $C = 0$. We go back and look at the $\boldsymbol{e}_2$-component of the linearized Euler equation in (6.6-8): Since we are considering a right-traveling wave with "velocity" $\frac{dx_1}{dx_2} = \sqrt{m^2-1}$, we have $\partial_2 \eta = -\sqrt{m^2-1}\,\partial_1 \eta$, and (6.6-8) becomes

$$\rho_0 u_0 \partial_1 v_2 - c^2\sqrt{m^2-1}\,\partial_1 \eta = 0.$$

Integrating with respect to x_1 and using $v_2 = 0$ for $x_1 < \sqrt{m^2-1}\,x_2$, we get

$$\rho_0 u_0 v_2 = c^2\sqrt{m^2-1}\,\eta,$$

and substituting (6.6-10) into the right-hand side gives

$$(6.6\text{-}11) \qquad v_2 = u_0\{h'(x_1 - \sqrt{m^2-1}\,x_2) + C\}.$$

Compatibility with (6.6-7) on $x_2 = 0$ gives $C = 0$. In summary, our results for η and v_2 in R are:

$$(6.6\text{-}12) \qquad \eta \;=\; \frac{\rho_0 m^2}{\sqrt{m^2-1}}h'(x_1 - \sqrt{m^2-1}\,x_2),$$

$$(6.6\text{-}13) \qquad v_2 \;=\; u_0 h'(x_1 - \sqrt{m^2-1}\,x_2).$$

Streamlines $x_2 = x_2(x_1)$ satisfy

$$\frac{dx_2}{dx_1} = \frac{v_2}{u_0 + v_1}.$$

For $u_0 \gg v_2$, we replace the right-hand side by $\frac{v_2}{u_0}$ and then substitute (6.6-13) for v_2. We get

$$\frac{dx_2}{dx_1} = h'(x_1 - \sqrt{m^2-1}\,x_2).$$

In the limit of small h', the asymptotic solutions of this ODE are

$$(6.6\text{-}14) \qquad x_2 = y + h(x_1 - \sqrt{m^2-1}\,y),$$

where y is the initial elevation of the streamline before it enters the wake. The streamlines are depicted in Figure 6.10.

For $h'(0^+) > 0$, the "leading edge" of the wake, $x_1 = \sqrt{m^2-1}\,x_2$, is *not* a wavefront with η and $\boldsymbol{v}$ zero along it. To make the observed kinks in the streamlines, we need the normal component of the velocity to jump, and

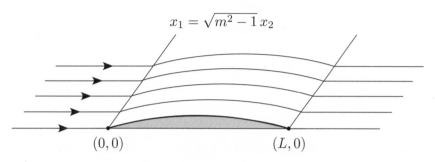

$$x_1 = \sqrt{m^2 - 1}\, x_2$$

$(0,0)$ $(L,0)$

Figure 6.10.

then a collateral density jump must follow in order to preserve the normal component of mass flux. In fact, the jump in density as you enter the wake is

$$[\eta] = \frac{\rho_0 m^2}{\sqrt{m^2 - 1}} h'(0^+).$$

For $h'(0^+) > 0$, this is *positive*, as we know it should be from our analysis of shocks in Problem 6.5. For the "trailing edge" shock along $x_1 = L + \sqrt{m^2 - 1}\, x_2$, we have

$$[\eta] = -\frac{\rho_0 m^2}{\sqrt{m^2 - 1}} h'(L^-),$$

which is also positive if $h'(L^-) < 0$.

Incompressible fluid. The idea is to look at flows for which changes in material volumes can be neglected. Recall that the volume $V(t)$ of a material region $R(t)$ has time rate of change

$$\dot{V} = \int_{R(t)} \boldsymbol{\nabla} \cdot \boldsymbol{u}\, d\boldsymbol{x}.$$

If $\dot{V} = 0$ for all $R(t)$, then

(6.32) $$\boldsymbol{\nabla} \cdot \boldsymbol{u} = 0$$

throughout the fluid. In this case, the mass conservation PDE (6.18) reduces to $\frac{D\rho}{Dt} = 0$. Here, assume the simplest situation, with ρ time-independent *and* uniform. In this case, it seems that the pressure $p = p(\rho)$ would be uniform as well, and there would be *no* acceleration of fluid particles. But this is *not* how nearly incompressible fluids really work. There is still a nonuniform pressure field, but its mode of determination is completely different.

Take the divergence of the Euler equation (6.15),

$$(6.33) \qquad\qquad \boldsymbol{u}_t + \boldsymbol{u} \cdot \boldsymbol{\nabla} \boldsymbol{u} = -\frac{1}{\rho} \boldsymbol{\nabla} p.$$

Since $\boldsymbol{\nabla} \cdot \boldsymbol{u} = 0$, this gives

$$(6.34) \qquad\qquad \Delta p = -\rho \boldsymbol{\nabla} \cdot (\boldsymbol{u} \cdot \boldsymbol{\nabla} \boldsymbol{u})$$

throughout the fluid. Suppose the fluid is confined to a *fixed* region R. Let $\boldsymbol{n}$ be outward unit normal on ∂R. Since fluid does not go through ∂R, $\boldsymbol{u} \cdot \boldsymbol{n} = 0$ on ∂R, and hence $\boldsymbol{u}_t \cdot \boldsymbol{n} = 0$ on ∂R. Taking the inner product of the Euler equation (6.33) with $\boldsymbol{n}$ now gives

$$(6.35) \qquad\qquad p_n = -\rho (\boldsymbol{u} \cdot \boldsymbol{\nabla} \boldsymbol{u}) \cdot \boldsymbol{n}$$

on ∂R. (6.34) and (6.35) constitute a Neumann boundary value problem for p, which determines p as a functional of $\boldsymbol{u}$, up to a constant. Hence, $\boldsymbol{\nabla} p$ is determined uniquely from $\boldsymbol{u}$, and the Euler equation (6.33) is now an evolution equation for $\boldsymbol{u}$ which upholds $\boldsymbol{\nabla} \cdot \boldsymbol{u} = 0$ for all t if $\boldsymbol{\nabla} \cdot \boldsymbol{u} = 0$ at $t = 0$.

One feature of the incompressible fluid model is immediately clear: The determination of pressure by the elliptic boundary value problem (6.34)–(6.35) means instantaneous propagation. Any change in $\boldsymbol{\nabla} \cdot (\boldsymbol{u} \cdot \boldsymbol{\nabla} \boldsymbol{u})$ leads to a pressure perturbation that is felt instantaneously throughout the fluid. At the present naive level of examination, there seems to be a big disconnect between the actual mechanics of compressible fluid and the incompressible model. The stage has been set for the first case study in scaling-based reduction.

Problem 6.7 (Two-dimensional vortex dynamics).   We have a two-dimensional incompressible flow $\boldsymbol{u}$ in a fixed region R of $\mathbb{R}^2$. The boundary ∂R is impenetrable, so $\boldsymbol{u} \cdot \boldsymbol{n} = 0$ on ∂R.

a) Show that there is a scalar field $\psi(\boldsymbol{x})$, called a *stream function*, such that

$$(6.7\text{-}1) \qquad\qquad u_1 = \partial_2 \psi, \quad u_2 = -\partial_1 \psi$$

in R. (Hint: Consider line integrals $\int_C -u_2 \, dx_1 + u_1 \, dx_2$.)

b) Show that $\psi(\boldsymbol{x})$ satisfies Poisson's equation

$$(6.7\text{-}2) \qquad\qquad \Delta \psi = -\omega$$

in R, where

$$(6.7\text{-}3) \qquad\qquad \omega := \partial_1 u_2 - \partial_2 u_1$$

is the *scalar vorticity* associated with the two-dimensional flow $\boldsymbol{u}$. Show that ψ is constant on each connected component of ∂R. In the case of simply connected R (for which ∂R is one simple closed curve), show that ψ is uniquely determined up to an additive constant (and so the velocity $\boldsymbol{u}$ is uniquely determined from ω).

c) We determine $\boldsymbol{u}$ as a functional of ω for the case where $R = \mathbb{R}^2$ and $\boldsymbol{u} \to 0$ at ∞. First, find ψ and then $\boldsymbol{u}$ for a point vortex with $\omega(\boldsymbol{x}) = \Gamma\delta(\boldsymbol{x})$; then apply superposition to get the flow $\boldsymbol{u}$ for continuous $\omega(\boldsymbol{x})$.

d) The scalar vorticity ω defined in (6.7-3) actually merits the title of "scalar": Show that in two-dimensional incompressible flows of ideal fluid, ω is indeed a convected scalar. Explain how the time derivative ω_t at any point in a simply connected R is uniquely determined from current values of ω throughout R.

Solution.

a) Let $\boldsymbol{x}_0$ be a fixed point in R and $\boldsymbol{x}$ any other (arbitrary) point in R, and let C be a curve in R from $\boldsymbol{x}_0$ to $\boldsymbol{x}$. We claim that the line integral $\int_C -u_2\,dx_1 + u_1\,dx_2$ is independent of C. Let C and C' be two curves in R from $\boldsymbol{x}_0$ to $\boldsymbol{x}$. Then $C - C'$ is a simple closed curve or a union of simple closed curves as depicted in Figure 6.11. Let S be the portion of R enclosed by one of these simple closed curves Γ. Then by incompressibility and the

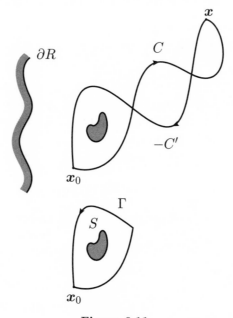

Figure 6.11.

divergence theorem,

$$(6.7\text{-}4) \qquad 0 = \int_S (\partial_1 u_1 + \partial_2 u_2)\, dx_1 dx_2 = \int_{\partial S} -u_2\, dx_1 + u_1\, dx_2.$$

If the whole interior of Γ is in R, then $\partial S = \Gamma$ (or $-\Gamma$), and $\int_\Gamma -u_2\, dx_1 + u_1\, dx_2 = 0$. If Γ surrounds a "hole" in R, as depicted in Figure 6.11, one component of ∂S is the boundary of the hole. On the hole boundary, $\boldsymbol{u} \cdot \boldsymbol{n}\, ds = (u_1\boldsymbol{e}_1 + u_2\boldsymbol{e}_2) \cdot (dx_2\boldsymbol{e}_1 - dx_1\boldsymbol{e}_2) = -u_2 dx_1 + u_1 dx_2 = 0$; so in the right-hand side of (6.7-4), the "hole boundary" contribution vanishes, and again $\int_\Gamma -u_2\, dx_1 + u_1\, dx_2 = 0$. In summary,

$$\int_C -u_2\, dx_1 + u_1\, dx_2 = \int_{C'} -u_2\, dx_1 + u_1\, dx_2.$$

The common value of all the line integrals $\int_C -u_2\, dx_1 + u_1\, dx_2$ for any curve C from $\boldsymbol{x}_0$ to $\boldsymbol{x}$ is some function $\psi(\boldsymbol{x})$ of $\boldsymbol{x}$. Now notice that

$$\psi(\boldsymbol{x} - h\boldsymbol{e}_1) - \psi(\boldsymbol{x}) = \int_0^h -u_2(\boldsymbol{x} + s\boldsymbol{e}_1)\, ds,$$

and in the limit $h \to 0$ we find $\partial_1 \psi(\boldsymbol{x}) = -u_2(\boldsymbol{x})$. Similarly, $\partial_2 \psi(\boldsymbol{x}) = u_1(\boldsymbol{x})$.

b) From (6.7-3) and (6.7-1) we calculate

$$\omega := \partial_1 u_2 - \partial_2 u_1 = \partial_1(-\partial_1 \psi) - \partial_2(\partial_2 \psi) = -\Delta \psi$$

in R. The no penetration boundary condition $\boldsymbol{u} \cdot \boldsymbol{n} = 0$ on ∂R can be rewritten as

$$0 = \boldsymbol{u} \cdot \boldsymbol{n} = \partial_2 \psi n_1 - \partial_1 \psi n_2 = (\partial_1 \psi \boldsymbol{e}_1 + \partial_2 \psi \boldsymbol{e}_2) \cdot (-n_2\boldsymbol{e}_1 + n_1\boldsymbol{e}_2) = \boldsymbol{\nabla} \psi \cdot \boldsymbol{\tau},$$

where $\boldsymbol{\tau} := -n_2\boldsymbol{e}_1 + n_1\boldsymbol{e}_2$ is a unit tangent of ∂R. Since the tangential derivative of ψ vanishes on ∂R, ψ is constant along each connected component of ∂R.

Now take simply connected R, with $\psi = \psi_0 = $ constant on ∂R. The solution of Poisson's equation (6.7-2) subject to the boundary condition $\psi = \psi_0$ on ∂R is unique: The difference η of two solutions is harmonic in R, with zero boundary condition on ∂R. With $\Delta \eta = 0$ in R, we have

$$\boldsymbol{\nabla} \cdot (\eta \boldsymbol{\nabla} \eta) = |\boldsymbol{\nabla} \eta|^2,$$

and integration over R and use of the divergence theorem gives

$$\int_{\partial R} \eta \partial_n \eta\, ds = \int_R |\boldsymbol{\nabla} \eta|^2\, d\boldsymbol{x}.$$

The line integral on the left-hand side is zero because $\eta = 0$ on ∂R. Hence $\boldsymbol{\nabla} \eta = 0$ in R, so η is constant, and this constant is zero by the zero boundary condition. In summary, ψ is uniquely determined up to the constant boundary value ψ_0. The velocity components obtained by differentiating ψ are therefore unique.

c) The relevant solution of Poisson's equation $\Delta\psi = -\Gamma\delta(\boldsymbol{x})$ is

(6.7-5) $$\psi = -\frac{\Gamma}{2\pi}\log r = -\frac{\Gamma}{4\pi}\log(x_1^2 + x_2^2).$$

We have $\psi_r \to 0$ as $r \to \infty$, so $\boldsymbol{u} \to 0$ at ∞. The velocity field is

$$\boldsymbol{u} = \partial_2\psi\boldsymbol{e}_1 - \partial_1\psi\boldsymbol{e}_2 = \frac{\Gamma}{2\pi}\frac{-x_2\boldsymbol{e}_1 + x_1\boldsymbol{e}_2}{x_1^2 + x_2^2}.$$

We have seen this point vortex flow before, in Figure 1.6. For continuous vorticity $\omega(\boldsymbol{x})$, the velocity field is given by the superposition

(6.7-6) $$\boldsymbol{u}(\boldsymbol{x}) = \frac{1}{2\pi}\int_{\mathbb{R}^2}\frac{-(x_2 - x_2')\boldsymbol{e}_1 + (x_1 - x_1')\boldsymbol{e}_2}{(x_1 - x_1')^2 + (x_2 - x_2')^2}\omega(\boldsymbol{x}')\,d\boldsymbol{x}'.$$

d) The convected scalar character of ω is an easy special case of Problem 6.4: Taking $u_3 = 0$ and u_1, u_2 independent of x_3, the *vector* vorticity $\omega = \nabla \times \boldsymbol{u}$ reduces to $(\partial_1 u_2 - \partial_2 u_1)\boldsymbol{e}_3 = \omega\boldsymbol{e}_3$, and the only component of the PDE (6.4-2) is the $\boldsymbol{e}_3$-component, which reduces to

(6.7-7) $$\omega_t + (\boldsymbol{u} \cdot \nabla)\omega = 0$$

in the incompressible case, where $\nabla \cdot \boldsymbol{u} = 0$. For two-dimensional incompressible flow of ideal fluid in our simply connected region R, recall that the velocity field at any point in R is uniquely determined from the vorticity field ω throughout R. In this way, (6.7-7) becomes a closed evolution equation for ω in R.

Problem 6.8 (The "ground effect").   A vorticity field $\omega(\boldsymbol{x}, t)$ in a region R of $\mathbb{R}^2$ satisfies the vortex dynamics set forth in Problem 6.7.

a) Show that $x_1\omega$ is locally convected and identify its flux $\boldsymbol{f}_1$. By evoking isotropy of space, $x_2\omega$ is locally conserved too. Identify its flux $\boldsymbol{f}_2$. Be careful here: Simply swapping the $\boldsymbol{e}_1$ and $\boldsymbol{e}_2$ axes reverses the orientation, and this is not quite what you want.

b) The center of vorticity is defined as

(6.8-1) $$\boldsymbol{X} := \frac{\int_R \boldsymbol{x}\omega\,d\boldsymbol{x}}{\int_R \omega\,d\boldsymbol{x}}.$$

Represent the velocity components $\dot{X}_1$ and $\dot{X}_2$ as line integrals over ∂R.

c) Specialize the line integral formulas for $\dot{X}_1$ and $\dot{X}_2$ to flows in the upper half plane $x_2 > 0$, with the x_1-axis being an impervious wall. In particular, show that $\dot{X}_1$ has the same sign as the total circulation $\Gamma := \int_{x_2>0}\omega\,d\boldsymbol{x}$, and that $\dot{X}_2 = 0$. There are traveling vortex solutions: In a frame of reference traveling at some velocity $v\boldsymbol{e}_1$, the vorticity is time-independent, and its support R is some finite region above the x_1-axis as depicted in Figure 6.12. The velocity $\boldsymbol{u}$ asymptotes to zero at ∞. Determine the horizontal velocity v as a functional of the vorticity. Evaluate v in the limit of a point vortex of circulation Γ at a distance a above the x_1-axis.

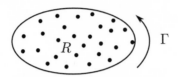

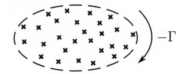

Figure 6.12.

Solution.

a) We have

$$(x_1\omega)_t + x_1\boldsymbol{u} \cdot \boldsymbol{\nabla}\omega = 0,$$

or

(6.8-2) $$(x_1\omega)_t + \boldsymbol{\nabla} \cdot (x_1\omega\boldsymbol{u}) - \omega u_1 = 0.$$

Due to the "extra" term $-\omega u_1$, we see that $x_1\omega$ is not simply a convected density in the flow $\boldsymbol{u}$. It remains to express ωu_1 as a divergence:

$$\omega u_1 = (\partial_1 u_2 - \partial_2 u_1)u_1 = u_1\partial_1 u_2 - \partial_2(\tfrac{1}{2}u_1^2)$$

$$= \partial_1(u_1 u_2) - u_2\partial_1 u_1 - \partial_2(\tfrac{1}{2}u_1^2)$$

$$= \partial_1(u_1 u_2) + u_2\partial_2 u_2 - \partial_2(\tfrac{1}{2}u_1^2)$$

$$= \partial_1(u_1 u_2) - \partial_2(\tfrac{1}{2}u_1^2 - \tfrac{1}{2}u_2^2).$$

The third equality uses incompressibility, $\partial_1 u_1 = -\partial_2 u_2$. Hence, (6.8-2) becomes

(6.8-3) $$(x_1\omega)_t + \boldsymbol{\nabla} \cdot \boldsymbol{f}_1 = 0,$$

where $\boldsymbol{f}_1$ is the flux

(6.8-4) $$\boldsymbol{f}_1 = x_1\omega\boldsymbol{u} - u_1 u_2\boldsymbol{e}_1 + \frac{1}{2}(u_1^2 - u_2^2)\boldsymbol{e}_2.$$

To find the flux $\boldsymbol{f}_2$ of $x_2\omega$, we do a $\frac{\pi}{2}$ rotation, so we have the replacements $x_1 \to x_2$, $x_2 \to -x_1$, $u_1 \to u_2$, $u_2 \to -u_1$ and $\boldsymbol{e}_1 \to \boldsymbol{e}_2$, $\boldsymbol{e}_2 \to -\boldsymbol{e}_1$. Hence, the flux of $x_2\omega$ is

(6.8-5) $$\boldsymbol{f}_2 = x_2\omega\boldsymbol{u} + \frac{1}{2}(u_1^2 - u_2^2)\boldsymbol{e}_1 + u_1u_2\boldsymbol{e}_2.$$

b) In (6.8-1), the total circulation

$$\Gamma := \int_R \omega\,d\boldsymbol{x} = \int_{\partial R} \boldsymbol{u} \cdot d\boldsymbol{x}$$

is a constant independent of time, so time differentiation of (6.8-1) gives

$$\dot{\boldsymbol{X}} = \frac{1}{\Gamma} \int_R \boldsymbol{x}\omega_t\,d\boldsymbol{x}.$$

Look at the $\boldsymbol{e}_1$-component:

$$\dot{X}_1 = \frac{1}{\Gamma} \int_R (x_1\omega)_t\,d\boldsymbol{x} = -\frac{1}{\Gamma} \int_R \boldsymbol{\nabla} \cdot \boldsymbol{f}_1\,d\boldsymbol{x} = -\frac{1}{\Gamma} \int_{\partial R} \boldsymbol{f}_1 \cdot \boldsymbol{n}\,ds.$$

Here, we have used (6.8-3). Inserting (6.8-4) for $\boldsymbol{f}_1$, we get

(6.8-6) $$\dot{X}_1 = \frac{1}{\Gamma} \int_{\partial R} \{u_1u_2\boldsymbol{e}_1 - \frac{1}{2}(u_1^2 - u_2^2)\boldsymbol{e}_2\} \cdot \boldsymbol{n}\,ds,$$

where we have used $\boldsymbol{u} \cdot \boldsymbol{n} = 0$ on ∂R. Similarly,

(6.8-7) $$\dot{X}_2 = -\frac{1}{\Gamma} \int_{\partial R} \{\frac{1}{2}(u_1^2 - u_2^2)\boldsymbol{e}_1 + u_1u_2\boldsymbol{e}_2\} \cdot \boldsymbol{n}\,ds.$$

c) With $\boldsymbol{n} = -\boldsymbol{e}_2$ and $u_2 = 0$ on $x_2 = 0$, (6.8-6) and (6.8-7) reduce to

(6.8-8) $$\dot{X}_1 = \frac{1}{2\Gamma} \int_{-\infty}^{\infty} u_1^2(x_1, 0)\,dx_1, \quad \dot{X}_2 = 0.$$

Clearly $\dot{X}_1$ has the same sign as the circulation Γ.

Our next task is to evaluate $u_1(x_1, 0)$ in terms of ω, so that we get $\dot{X}_1$ as a functional of ω. If the flow were in unbounded $\mathbb{R}^2$, we would have the integral representation of $\boldsymbol{u}$ as in (6.7-6). This flow generally does not satisfy $u_2 = 0$ on $x_2 = 0$. But if we do an *odd* extension of the vorticity field about the x_1-axis as indicated in Figure 6.12, the resulting flow due to the original vortex and the "image" vortex nicely satisfies $u_2 = 0$ on $x_2 = 0$. The effect on $u_1(x_1, 0)$ is to give twice the unbounded $\mathbb{R}^2$ value as follows from (6.7-6); that is,

(6.8-9) $$u_1(x_1, 0) = \frac{1}{\pi} \int_{x_2 > 0} \frac{x_2'}{(x_1 - x_1')^2 + x_2'^2}\omega(x_1', x_2')\,dx_1dx_2.$$

For the point vortex with $\omega := \Gamma\delta(\boldsymbol{x} - a\boldsymbol{e}_2)$, we get

$$u_1(x_1, 0) = \frac{\Gamma}{\pi} \frac{a}{x_1^2 + a^2},$$

and then (6.8-8) gives

$$v = \frac{\Gamma a^2}{2\pi^2} \int_{-\infty}^{\infty} \frac{dx_1}{(a^2 + x_1^2)^2} = \frac{\Gamma}{4\pi a}.$$

Notice that the flow due to the image vortex at $x = -ae_2$ evaluated at $x = ae_2$ gives $\frac{\Gamma}{4\pi a}e_1$.

Problem 6.9 (Elliptical vortices).

a) Let u be the incompressible flow with uniform vorticity ω_0 inside the ellipse

(6.9-1) $$\frac{x_1^2}{a^2} + \frac{x_2^2}{b^2} = 1,$$

zero vorticity outside, and $u \to 0$ as $r \to \infty$. Find u *inside* the ellipse. (Hint: The stream function of the flow is a harmonic function of the complex variable $z = x_1 + ix_2$ in the exterior of the ellipse. Assuming $a > b$, we map the exterior of the unit circle in the $\omega := \omega_1 + i\omega_2$ plane into the exterior of the ellipse in the z-plane by

(6.9-2) $$z = \frac{a+b}{2}\omega + \frac{a-b}{2}\frac{1}{\omega}.$$

The stream function ψ expressed as a function of ω is a harmonic function of ω in $|\omega| > 1$.)

b) We have two-dimensional incompressible ideal fluid in $\mathbb{R}^2$, and the initial velocity field at time 0 is the flow in Part a. How does the vorticity field evolve in time?

c) Next, we look for a time-independent two-dimensional flow of incompressible ideal fluid that has uniform vorticity ω_0 inside the ellipse (6.9-1) and zero vorticity outside, but with the velocity field asymptoting to the *strain flow* with stream function

(6.9-3) $$\psi_s = \frac{\gamma}{2}(x_1^2 - x_2^2)$$

as $r \to \infty$. Here, $\gamma > 0$ is a positive constant. Figure 6.13 depicts the streamlines of the strain flow. Determine the relationship between the ellipse aspect ratio $\frac{a}{b}$ and $\frac{\gamma}{\omega_0}$ so that such a flow is in fact realizable.

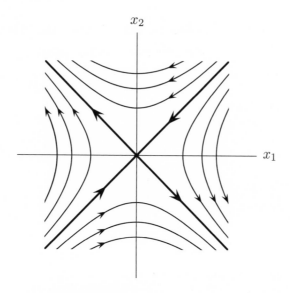

$$x_2$$

$$x_1$$

Figure 6.13.

Solution.

a) The stream function ψ satisfies

(6.9-4) $$\Delta \psi = \left[\begin{array}{cc} -\omega_0 & \text{inside ellipse,} \\ 0 & \text{outside.} \end{array} \right.$$

The simplest particular solution of $\Delta \psi = -\omega_0$ is

(6.9-5) $$\psi = -\frac{1}{2}(\alpha x_1^2 + \beta x_2^2),$$

where the constants α and β satisfy

(6.9-6) $$\alpha + \beta = \omega_0.$$

We assume that the solution for ψ inside the ellipse *is* (6.9-5), and that an additive harmonic function is unnecessary. The stream function ψ continues into the exterior as a harmonic function, with values and normal derivative on the ellipse inherited from the interior solution (6.9-5). (ψ and ψ_n are *continuous* across the ellipse.) The technical implementation of continuity conditions is simplest using polar coordinates r, ϑ of the ω-plane. Setting $\omega = r e^{i\vartheta}$ in (6.9-2), we find

(6.9-7)
$$x_1 = \left(\frac{a+b}{2}r + \frac{a-b}{2}\frac{1}{r} \right) \cos \vartheta,$$
$$x_2 = \left(\frac{a+b}{2}r - \frac{a-b}{2}\frac{1}{r} \right) \sin \vartheta.$$

Setting $r = 1$ gives parametric equations of the ellipse (6.9-1) in the z-plane. $r > 1$ corresponds to the exterior of the ellipse and $\sqrt{\frac{a-b}{a+b}} < r < 1$ to the interior. Substituting x_1 and x_2 from (6.9-7) with $r = 1$ into (6.9-5) gives the boundary values of ψ on $r = 1$ in the ω-plane,

$$(6.9\text{-}8) \qquad \psi(r = 1, \vartheta) = -\frac{1}{4}(\alpha a^2 + \beta b^2) - \frac{1}{4}(\alpha a^2 - \beta b^2)\cos 2\vartheta.$$

We also calculate ψ_r along $r = 1$: From (6.9-5),

$$\psi_r = -\alpha x_1 \partial_r x_1 - \beta x_2 \partial_r x_2,$$

with x_1 and x_2 as in (6.9-7). Evaluation on $r = 1$ gives

$$(6.9\text{-}9) \qquad \psi_r(1, \vartheta) = -\frac{ab}{2}(\alpha + \beta) - \frac{ab}{2}(\alpha - \beta)\cos 2\vartheta.$$

The harmonic functions ψ in $r > 1$ which take values (6.9-8) on $r = 1$ and are such that $\psi_r \to 0$ as $r \to \infty$ (remember that $\boldsymbol{u} = 0$ at ∞) are

$$(6.9\text{-}10) \qquad \psi = D\log r - \frac{1}{4}(\alpha a^2 + \beta b^2) - \frac{1}{4}(\alpha a^2 - \beta b^2)\frac{\cos 2\vartheta}{r^2},$$

where D is an arbitrary constant. It follows that

$$\psi_r(1, \vartheta) = D + \frac{1}{2}(\alpha a^2 - \beta b^2)\cos 2\vartheta,$$

and consistency with (6.9-9) forces

$$(6.9\text{-}11) \qquad\qquad\qquad D = -\frac{ab}{2}(\alpha + \beta),$$

$$(6.9\text{-}12) \qquad\qquad\qquad \alpha a - \beta b = 0.$$

(6.9-6) and (6.9-12) constitute a *pair* of equations for α and β, giving $\alpha = \frac{b}{a+b}\omega_0$, $\beta = \frac{a}{a+b}\omega_0$. The solution (6.9-5) inside the ellipse is

$$(6.9\text{-}13) \qquad\qquad \psi = -\frac{1}{2}\frac{\omega_0}{a+b}(bx_1^2 + ax_2^2),$$

and the interior velocity field is

$$(6.9\text{-}14) \qquad \boldsymbol{u} = \partial_2\psi\boldsymbol{e}_1 - \partial_1\psi\boldsymbol{e}_2 = -\frac{\omega_0}{a+b}(ax_2\boldsymbol{e}_1 - bx_1\boldsymbol{e}_2).$$

We *can* pursue the details of the exterior flow, but that is unnecessary.

b) First, our "vortex patch" of uniform vorticity ω_0 is a material region, as shown in Problem 6.7. The flow (6.9-14) inside the ellipse is *linear* in $\boldsymbol{x}$. Since linear flow maps carry ellipses into ellipses, we expect that our vortex patch remains an ellipse for all time. It turns out that with no flow at ∞, our ellipse rotates without change of shape. Here is a simple way to recover the angular velocity of rotation without tedious exercise of linear algebra: In a frame of reference that rotates at angular velocity Ω, the interior stream function is

$$(6.9\text{-}15) \qquad \psi' = -\frac{1}{2}\frac{\omega_0}{a+b}(bx_1^2 + ax_2^2) + \frac{\Omega}{2}(x_1^2 + x_2^2).$$

Note that the contribution to velocity generated by $\frac{\Omega}{2}(x_1^2 + x_2^2)$ is $\Omega x_2 e_1 - \Omega x_1 e_2$, which is solid body rotation with angular velocity $-\Omega$. If Ω is in fact equal to the rotation rate of our ellipse, the flow in the rotating frame would be steady, and our ellipse would be a streamline with ψ *constant* on it. Setting $x_1 = a\cos\varphi$ and $x_2 = b\sin\varphi$ (parametric equations of the ellipse) in (6.9-15), we have

$$x' = \frac{1}{2}\left(\Omega - \frac{\omega_0 b}{a+b}\right)a^2\cos^2\varphi + \frac{1}{2}\left(\Omega - \frac{\omega_0 a}{a+b}\right)b^2\sin^2\varphi.$$

The right-hand side reduces to a constant independent of φ if

$$\left(\Omega - \frac{\omega_0 b}{a+b}\right)a^2 = \left(\Omega - \frac{\omega_0 a}{a+b}\right)b^2,$$

from which it follows that

$$\Omega = \frac{\omega_0 ab}{(a+b)^2}.$$

c) The stream function ψ_s in (6.9-3) associated with the strain flow is harmonic, so we can add it to the "vortex" stream function in Part b and preserve the same vortex patch (with vorticity ω_0 inside the ellipse). The superposition of vortex and strain flows asymptotes to the strain flow at infinity. Notice that this superposition of velocity fields is *still linear in* x *inside the ellipse*. Evidently, the vortex patch will still evolve in time as an ellipse, but its axes and orientation undergo some evolution that is generally more complicated than simple rotation. This ellipse dynamics can be analyzed using linear algebra to derive ODEs for the ellipse parameters. Our problem as stated is less ambitious—all we want are the fixed points of this ellipse dynamics. This lesser problem is easily done: Simply arrange

$$\psi' := -\frac{1}{2}\frac{\omega_0}{a+b}(bx_1^2 + ax_2^2) + \frac{\gamma}{2}(x_1^2 - x_2^2)$$

to be constant on the ellipse. This is a repeat of the analysis in Part b, and we quickly determine the relationship between the aspect ratio $R := \frac{a}{b}$ and $\frac{\gamma}{\omega_0}$ to be

$$(6.9\text{-}16) \qquad \frac{\gamma}{\omega_0} = \frac{R(R-1)}{(R+1)(R^2+1)}.$$

Figure 6.14 is the graph of $\frac{\gamma}{\omega_0}$ versus $\frac{a}{b}$.

If $\frac{\gamma}{\omega_0} < 1.5$, we get two possible R's. It turns out that the elliptical vortex with the higher aspect ratio R_+ is unstable, even within the space of

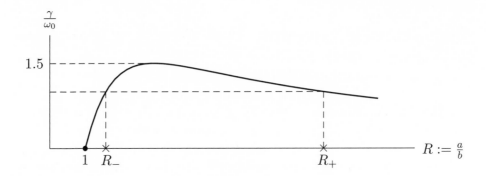

Figure 6.14.

elliptical configurations, while the smaller aspect ratio (R_-) ellipse is stable. If $\frac{\gamma}{\omega_0} > 1.5$, the strain flow is too powerful for a time-independent vortex, and the eventual fate of an elliptical vortex is its extrusion into a long "cigar" parallel to the *outgoing* strain axis $x_2 = -x_1$ in Figure 6.13.

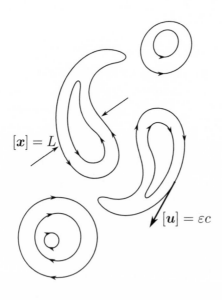

Figure 6.15.

The incompressible limit of compressible flow. In many of the fluid flows of "everyday life", such as the air flow around a VW bug on the highway, the fluid particles move much more slowly than the speed of sound. To describe such flows, the unit of flow velocity u is taken to be εc, where c is the speed of sound and ε is a dimensionless gauge parameter. A limit with $\varepsilon \to 0$ is to be considered. The smallness of flow velocity by itself does not single out the incompressible limit; after all, the velocities of fluid particles in a small-amplitude sound wave are much smaller than the speed of sound. One needs to consider the characteristic time associated with a flow in relation to its characteristic length. Let L be a characteristic length of the flow. For instance, L could be the size of some eddies, as depicted in Figure 6.15. The time required for a fluid particle to circulate around one of the eddies is on the order of $\frac{L}{\varepsilon c}$. This is the unit of time appropriate for the incompressible limit. Notice that it is much longer than $\frac{L}{c}$, the time it takes for sound to propagate across one of the eddies. The above units of length, time and velocity are tabulated in the following table:

(6.36)

Variable	$\boldsymbol{x}$	t	$\boldsymbol{u}$	$\eta := \rho - \rho_0$
Unit	L	$\frac{L}{\varepsilon c}$	εc	$\varepsilon^2 \rho_0$

The scaling of the density perturbation

$$\text{(6.37)} \qquad \eta := \rho - \rho_0$$

from a "rest density" ρ_0 is easily identified from the Euler equation (6.19): The characteristic acceleration of fluid particles is the unit of length divided by the unit of time squared,

$$\text{(6.38)} \qquad \frac{L}{(L/\varepsilon c)^2} = \frac{\varepsilon^2 c^2}{L}.$$

Now look at the pressure force per unit mass when $\frac{|\eta|}{\rho_0} \ll 1$:

$$\text{(6.39)} \qquad \frac{1}{\rho}\boldsymbol{\nabla}p = \frac{1}{\rho}p'(\rho)\boldsymbol{\nabla}\rho \sim \frac{c^2}{\rho_0}\boldsymbol{\nabla}\eta.$$

Let $[\eta]$ denote the unit of η, to be determined. The pressure force in (6.39) has magnitude

$$\text{(6.40)} \qquad \frac{c^2}{\rho_0 L}[\eta].$$

Since the pressure force per unit mass balances acceleration, equality of (6.38) and (6.40) gives

$$[\eta] = \varepsilon^2 \rho_0.$$

This unit of η is entered into the scaling table (6.36).

From (6.18) and (6.19) it follows that the dimensionless equations for η and $\boldsymbol{u}$ are

$$\text{(6.41)} \qquad \varepsilon^2(\eta_t + \boldsymbol{u} \cdot \boldsymbol{\nabla}\eta) + (1 + \varepsilon^2\eta)\boldsymbol{\nabla} \cdot \boldsymbol{u} = 0,$$

$$\text{(6.42)} \qquad \boldsymbol{u}_t + \boldsymbol{u} \cdot \boldsymbol{\nabla}\boldsymbol{u} = -\frac{p'(\rho(1 + \varepsilon^2\eta))}{c^2(1 + \varepsilon^2\eta)}\boldsymbol{\nabla}\eta.$$

We discuss the transcription of (6.18)–(6.19) into (6.41)–(6.42). To transcribe $\rho_t = \eta_t$, divide the unit of η by the unit of t, so that the dimensional η_t is given in terms of the dimensionless η_t by $(\varepsilon^2\eta_0)(\varepsilon c/L)\eta_t$. Similarly, in $\boldsymbol{u} \cdot \boldsymbol{\nabla}\rho = \boldsymbol{u} \cdot \boldsymbol{\nabla}\eta$, multiply the units of $\boldsymbol{u}$ and η and divide by the unit of length. Hence, dimensional $\boldsymbol{u} \cdot \boldsymbol{\nabla}\rho$ in terms of dimensionless $\boldsymbol{u} \cdot \boldsymbol{\nabla}\eta$ is $(\varepsilon c)(\varepsilon^2\rho_0)(1/L)\boldsymbol{u} \cdot \boldsymbol{\nabla}\eta$. Similarly, $\rho\boldsymbol{\nabla} \cdot \boldsymbol{u} = (\rho_0 + \eta)\boldsymbol{\nabla} \cdot \boldsymbol{u}$ is replaced by

$(\rho_0)(\varepsilon c)(1/L)(1+\varepsilon^2\eta)\boldsymbol{\nabla}\cdot\boldsymbol{u}$. In summary, the dimensional mass conservation PDE is transcribed into its dimensionless form

$$(\varepsilon^2\eta_0)\left(\frac{\varepsilon c}{L}\right)\eta_t + (\varepsilon c)(\varepsilon^2\rho_0)\left(\frac{1}{L}\right)\boldsymbol{u}\cdot\boldsymbol{\nabla}\eta + (\rho_0)(\varepsilon c)\left(\frac{1}{L}\right)(1+\varepsilon^2\eta)\boldsymbol{\nabla}\cdot\boldsymbol{u} = 0.$$

Now comes cancellation of common factors, and what remains is (6.41). The dimensionless Euler equation (6.42) is derived by the same technique.

The incompressible limit of compressible flow is now elementary. The $\varepsilon \to 0$ limit of the mass conservation PDE (6.41) is simply

(6.43) $$\boldsymbol{\nabla}\cdot\boldsymbol{u} = 0,$$

as in the incompressible model. In the dimensionless Euler equation (6.42),

$$p'(\rho_0(1+\varepsilon^2\eta)) \to p'(\rho_0) = c^2$$

as $\varepsilon \to 0$, so the $\varepsilon \to 0$ limit of (6.42) is

(6.44) $$\boldsymbol{u}_t + \boldsymbol{u}\cdot\boldsymbol{\nabla}\boldsymbol{u} = -\boldsymbol{\nabla}\eta.$$

Restoring physical units to this reduced Euler equation gives

$$\left(\frac{1}{\varepsilon c}\right)\left(\frac{L}{\varepsilon c}\right)\boldsymbol{u}_t + \left(\frac{1}{\varepsilon c}\right)^2(L)\boldsymbol{u}\cdot\boldsymbol{\nabla}\boldsymbol{u} = -\left(\frac{1}{\varepsilon^2\rho_0}\right)(L)\boldsymbol{\nabla}\eta,$$

or

(6.45) $$\boldsymbol{u}_t + \boldsymbol{u}\cdot\boldsymbol{\nabla}\boldsymbol{u} = -\frac{c^2}{\rho_0}\boldsymbol{\nabla}\eta.$$

Hence, the pressure fluctuations in the "incompressible" model are in fact proportional to density fluctuations. From a naive point of view, *retention* of density fluctuations in the mass conservation PDE and their *deletion* from the Euler equation might seem inconsistent. But under the precise $\varepsilon \to 0$ limit process presented here, that is exactly what happens.

Problem 6.10 (Buoyancy force).   This problem is another example of scaling-based reduction. We have incompressible ideal fluid, but with density variations (such as salty and fresh water, or warm and cold water). In addition, the fluid is subject to a gravity force of $-g\boldsymbol{e}_3$ per unit mass.

a) Write down the PDE governing the density ρ, velocity $\boldsymbol{u}$ and pressure p.

b) Find the admissible pressure fields p for *hydrostatic equilibrium* ($\boldsymbol{u} \equiv 0$) of uniform fluid ($\rho \equiv \rho_0 =$ positive constant). Now suppose this hydrostatic equilibrium is perturbed by small nonuniformities of density. Let $\eta := \rho - \rho_0$ denote the density perturbation, and π the deviation of pressure from the hydrostatic pressure field. Formulate the PDE of Part a in terms of η and π

instead of ρ and p. Identify the *buoyancy force*, which is the force per unit mass that fluid with $\eta \neq 0$ would feel if the pressure were purely hydrostatic ($\pi = 0$).

c) We nondimensionalize: The scaling unit of density perturbation is $[\eta] = \varepsilon \rho_0$. Here, $\varepsilon > 0$ is a gauge parameter, and we will eventually consider a certain $\varepsilon \to 0$ limit. The scaling unit $[x]$ of x is given, presumably as a characteristic length associated with spatial variations of η. Now pick scaling units $[t]$, $[u]$, and $[\pi]$ for time t, velocity $\boldsymbol{u}$, and the pressure perturbation π so that (i) fluid particles move distances on the order of $[x]$ in times on the order of $[t]$, and (ii) the buoyancy force and force per unit mass due to the pressure perturbation (proportional to $\boldsymbol{\nabla}\pi$) have the same order of magnitude. Write down the dimensionless equations and take the $\varepsilon \to 0$ limit. Finally, restore physical units to the limit equations. This will be useful in the next problem.

Solution.

a) Since the flow is incompressible,

$$(6.10\text{-}1) \qquad\qquad \boldsymbol{\nabla} \cdot \boldsymbol{u} = 0$$

and the mass conservation PDE (6.8) reduces to

$$(6.10\text{-}2) \qquad\qquad \rho_t + (\boldsymbol{u} \cdot \boldsymbol{\nabla})\rho = 0$$

so that the density is a convected scalar. The Euler equation (6.15) modified by the gravity force is

$$(6.10\text{-}3) \qquad\qquad \boldsymbol{u}_t + (\boldsymbol{u} \cdot \boldsymbol{\nabla})\boldsymbol{u} = -\frac{1}{\rho}\boldsymbol{\nabla}p - g\boldsymbol{e}_3.$$

b) In a hydrostatic equilibrium with $\boldsymbol{u} \equiv 0$ and $\rho \equiv \rho_0$, the Euler equation (6.10-3) reduces to

$$\boldsymbol{\nabla}p = -g\rho_0 \boldsymbol{e}_3,$$

and the solution for p is

$$p = -g\rho_0 x_3$$

modulo an additive constant.

The mass conservation PDE (6.10-2) in terms of the density perturbation $\eta := \rho - \rho_0$ is

$$(6.10\text{-}4) \qquad\qquad \eta_t + \boldsymbol{u} \cdot \boldsymbol{\nabla}\eta = 0,$$

and the Euler equation (6.10-3) with the right-hand side expressed in terms of η and $\pi := p - (-g\rho_0 x_3)$ is

$$(6.10\text{-}5) \qquad\qquad \boldsymbol{u}_t + (\boldsymbol{u} \cdot \boldsymbol{\nabla})\boldsymbol{u} = -g\frac{\eta}{\rho_0 + \eta}\boldsymbol{e}_3 - \frac{1}{\rho_0 + \eta}\boldsymbol{\nabla}\pi.$$

The buoyancy force is the first term on the right-hand side, $-g\frac{\eta}{\rho_0+\eta}e_3$.

c) With $[\eta] = \varepsilon\rho_0$ and assuming the other scaling units $[x]$, $[t]$, $[u]$, $[\pi]$ to be given, the nondimensionalized versions of (6.10-4) and (6.10-5) are

(6.10-6) $$\frac{1}{[t]}\eta_t + \frac{[u]}{[x]}(\boldsymbol{u}\cdot\boldsymbol{\nabla})\eta = 0,$$

(6.10-7) $$\frac{[u]}{[t]}\boldsymbol{u}_t + \frac{[u]^2}{[x]}(\boldsymbol{u}\cdot\boldsymbol{\nabla})\boldsymbol{u} = -g\frac{\varepsilon\eta}{1+\varepsilon\eta}\boldsymbol{e}_3 - \frac{[\pi]}{[x]}\frac{1}{\rho_0}\frac{1}{(1+\varepsilon\eta)}\boldsymbol{\nabla}\pi.$$

The incompressibility equation (6.10-1) is invariant under scaling. If a fluid particle moves distance $[x]$ in time $[t]$ with velocity $[u]$, we have $[x] = [u][t]$, and (6.10-6) reduces to (6.10-4); that is to say, (6.10-4) is invariant under scalings with $[x] = [u][t]$. The Euler equation (6.10-7) becomes

(6.10-8) $$\frac{[u]^2}{[x]}\{\boldsymbol{u}_t + (\boldsymbol{u}\cdot\boldsymbol{\nabla})\boldsymbol{u}\} = -\frac{\varepsilon g\eta}{1+\varepsilon\eta}\boldsymbol{e}_3 - \frac{[\pi]}{[x]}\frac{1}{\rho_0}\frac{1}{(1+\varepsilon\eta)}\boldsymbol{\nabla}\pi$$

and, upon imposing the order-of-magnitude balance between all three terms, we deduce that

$$[u] = \sqrt{\varepsilon g[x]},$$
$$[\pi] = \varepsilon g\rho_0[x],$$

and finally

$$[t] = \frac{[x]}{[u]} = \sqrt{\frac{[x]}{\varepsilon g}}.$$

Given these scaling units, (6.10-8) reduces to

$$\boldsymbol{u}_t + (\boldsymbol{u}\cdot\boldsymbol{\nabla})\boldsymbol{u} = -\frac{\eta}{1+\varepsilon\eta}\boldsymbol{e}_3 - \frac{\boldsymbol{\nabla}\pi}{1+\varepsilon\eta}.$$

In summary, the dimensionless equations in the limit $\varepsilon \to 0$ are:

(6.10-9) $$\boldsymbol{\nabla}\cdot\boldsymbol{u} = 0,$$

(6.10-10) $$\eta_t + (\boldsymbol{u}\cdot\boldsymbol{\nabla})\eta = 0,$$

(6.10-11) $$\boldsymbol{u}_t + (\boldsymbol{u}\cdot\boldsymbol{\nabla})\boldsymbol{u} = -\eta\boldsymbol{e}_3 - \boldsymbol{\nabla}\pi.$$

Restoring physical units to these limit equations, the first two remain invariant and the last, the reduced Euler equation, becomes

(6.10-12) $$\boldsymbol{u}_t + (\boldsymbol{u}\cdot\boldsymbol{\nabla})\boldsymbol{u} = -\frac{g}{\rho_0}\eta\boldsymbol{e}_3 - \frac{1}{\rho_0}\boldsymbol{\nabla}\pi.$$

Problem 6.11 (Internal waves). We are going to apply the reduced model of buoyancy effects derived in Problem 6.10. In particular, we look at internal waves, which are depicted in Figure 6.16. In the far upstream limit

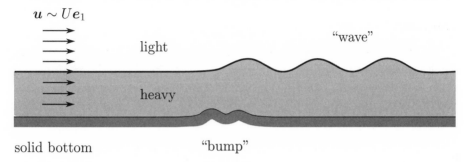

$$\boldsymbol{u} \sim U\boldsymbol{e}_1$$

light

"wave"

heavy

solid bottom

"bump"

Figure 6.16.

$x_1 \to -\infty$, $\boldsymbol{u}$ asymptotes to the uniform $\boldsymbol{u} = U\boldsymbol{e}_1$, where U is a positive constant, and the fluid is stably stratified with $\eta \to \eta_0(x_3)$, a decreasing positive function in $x_3 > 0$. In an interval of x_1 about $x_1 = 0$, there is a perturbation of the bottom topography from $x_3 = 0$, and we inquire into the possibility of an induced stationary wave of x_1-dependent density oscillations as $x_1 \to +\infty$. You have probably seen the wake of stationary waves on a swift river downstream from a submerged rock. That is essentially the same phenomenon.

a) First, some simplification of the equations (6.10-9), (6.10-10) and (6.10-12). We will relabel the vertical unit vector as $\boldsymbol{e}_2$ and we look for two-dimensional time-independent flows. That is, $\boldsymbol{u} = u_1\boldsymbol{e}_1 + u_2\boldsymbol{e}_2$, and η, u_1, u_2 and π are functions of x_1 and x_2 independent of t. We will deal with the incompressibility equation (6.10-9) by introducing a stream function $\psi(x_1, x_2)$ as we did in Problem 6.7. Show that η is a function of ψ. If $\boldsymbol{u} \to U\boldsymbol{e}_1$ and $\eta \to \eta_0(x_2)$, a given function in the upstream limit $x_1 \to -\infty$, what is your determination of η as a function of ψ? Next, derive a PDE for ψ. We are going to take the bottom to be $x_2 = 0$, and consider asymptotic stationary waves as $x_1 \to +\infty$. What is the boundary condition on ψ along $x_2 = 0$?

b) Formulate the linearized boundary value problem for the perturbation $\delta\psi$ of ψ from the uniform flow solution $\psi = Ux_2$. We will assume that $\delta\psi \to 0$ as $x_2 \to \infty$. Next, we seek stationary wave solutions of the form

$$(6.11\text{-}1) \qquad \delta\psi = e^{ikx_1}F(x_2).$$

Formulate the ODE eigenvalue problem for $F(x_2)$ and wavenumber k as a function of U. Attempt its explicit solution for

$$(6.11\text{-}2) \qquad \eta_0(x_2) = \begin{bmatrix} \eta_*, & 0 < x_2 < H, \\ 0, & x_2 > H. \end{bmatrix}$$

Here η_* is the drop in density as we go from heavy to light fluid in Figure 6.16, and H is the rest thickness of the heavy layer.

Solution.

a) In the time-independent case, (6.10-10) reduces to $(\boldsymbol{u} \cdot \boldsymbol{\nabla})\eta = 0$, so η is constant along integral curves of $\boldsymbol{u}$, which are the streamlines along which ψ is constant as well. Hence, $\eta = \eta(\psi)$, a function of ψ. From the upstream flow $\psi \to U x_2$, $\eta \to \eta_0(x_2)$ as $x_1 \to -\infty$, we deduce that η as a function of ψ is

$$(6.11\text{-}3) \qquad\qquad \eta = \eta_0 \left(\frac{\psi}{U} \right).$$

The PDE for ψ follows by taking the curl of the Euler equation (6.10-12) (with $\boldsymbol{e}_2$ in place of $\boldsymbol{e}_3$): Taking the x_2-derivative of the $\boldsymbol{e}_1$-component and subtracting the x_1-derivative of the $\boldsymbol{e}_2$-component results in

$$(6.11\text{-}4) \qquad\qquad (\boldsymbol{u} \cdot \boldsymbol{\nabla})\Delta\psi = \frac{g}{\rho_0}\partial_1 \eta$$

and, with η as in (6.11-3),

$$(6.11\text{-}5) \qquad\qquad (\boldsymbol{u} \cdot \boldsymbol{\nabla})\Delta\psi = \frac{g}{\rho_0 U}\eta_0' \left(\frac{\psi}{U} \right) \partial_1 \psi.$$

Since the bottom is $x_2 = 0$, (6.11-5) applies in $x_2 > 0$. The no penetration boundary condition on $x_2 = 0$ is $\psi(x_1,0) = \psi_0 = $ constant. Since the flow associated with a given stream function is found by differentiation, an additive constant has no physical meaning, and its value can be chosen for convenience. Hence, take $\psi_0 = 0$, and then $\psi(x_1,0) = 0$.

b) The linearized boundary value problem is

$$(6.11\text{-}6) \qquad\qquad \partial_1 \Delta\delta\psi = \frac{g}{\rho_0 U^2}\eta_0'(x_2)\partial_1 \delta\psi$$

in $x_2 > 0$, with zero boundary conditions on $x_2 = 0$ and as $x_2 \to \infty$. It follows from (6.11-6) that

$$\Delta\delta\psi - \frac{g}{\rho_0 U^2}\eta'(x_2)\delta\psi = f(x_2),$$

where $f(x_2)$ is an arbitrary function of x_2 which vanishes on $x_2 = 0$ and at ∞. What should $f(x_2)$ be? Recall that we are analyzing asymptotic stationary waves as $x_1 \to +\infty$, far downstream from the "bump" in Figure 6.16. Far upstream of the bump, we are supposed to have undisturbed uniform flow with $\delta\psi = 0$. This suggests taking $f(x_2) = 0$, so that

$$\delta\psi - \frac{g}{\rho_0 U^2}\eta_0'(x_2)\psi = 0$$

in $x_2 > 0$. Taking $\delta\psi$ in the separation-of-variables form (6.11-1), we find the ODE for F to be

$$F'' - \left\{ k^2 - \frac{g}{\rho_0 U^2} \eta_0'(x_2) \right\} F = 0$$

in $x_2 > 0$. For η_0 as in (6.11-2), we have $\eta_0'(x_2) = \eta_* \delta(x_2 - H)$, so

$$F'' - \left\{ k^2 - \frac{\eta_*}{\rho_0} \frac{g}{U^2} \delta(x_2 - H) \right\} F = 0.$$

We do the usual reformulation as

(6.11-7) $$F'' - k^2 F = 0$$

in $0 < x_2 < H$ and in $x_2 > H$, with jump conditions at $x_2 = H$:

(6.11-8)
$$[F] = 0,$$
$$[F'] = -\frac{\eta_*}{\rho_0} \frac{g}{U^2} F(x_2).$$

Finally, F satisfies zero boundary conditions on $x_2 = 0$ and at ∞. The solution of (6.11-7) which vanishes at $x_2 = 0$, ∞ and is continuous across $x_2 = H$ is

(6.11-9) $$F = \left[\begin{array}{ll} \sinh kx_2, & 0 < x_2 < H, \\[2ex] \sinh kL e^{-k(x_2 - L)}, & x_2 > H, \end{array} \right.$$

modulo a multiplicative constant. Substituting (6.11-9) into the derivative jump condition (second equation of (6.11-8)) yields a relationship between the wavenumber k and the upstream velocity U,

(6.11-10) $$\frac{kHe^{kH}}{\sinh kH} = \frac{\eta_*}{\rho_0} \frac{\delta H}{U^2}.$$

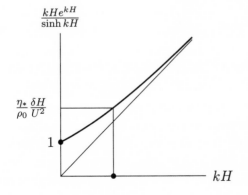

Figure 6.17.

Figure 6.17 depicts the graphical solution for k given U: It all depends on the dimensionless number $\frac{\eta_*}{\rho_0} \frac{\delta H}{U^2}$. If this is bigger than one, we have one real solution for the wavenumber k of the stationary wave. If this is less than one, there is no real solution. We can think of it in this way: In the rest frame of unperturbed fluid, U in (6.11-10) represents the plane speed of a traveling wave with wavenumber k. From (6.11-10) we see that the fastest speed $U_{\max} := \sqrt{\frac{\eta_*}{\rho_0} gH}$ is achieved in the long wave limit $kH \to 0$. As k increases, the wave speed decreases. If $U > U_{\max}$, there is *no* wave fast enough (with respect to the rest frame of the fluid) to "stand".

Guide to bibliography. Recommended references for this chapter are Batchelor [2], Chorin & Mardsen [4], Courant & Hilbert volume II [9], Feynman volume II [13], Landau & Lifshitz [17], Lighthill [18], Ockendon & Ockendon [20], Saffman [21], and Whitham [24].

The primary references for the basic derivation of fluid PDEs are Batchelor [2], Chorin & Mardsen [4], Feynman volume II [13], Landau & Lifshitz [17], and Ockendon & Ockendon [20]. Ockendon & Ockendon's treatment of the local mass and momentum conservation PDEs is closest to ours, especially in their preference for formulating integral forms of the conservation laws with respect to moving material volumes.

Bernoulli's theorem (the subject of Problem 6.3 in this text) is presented briefly in Chorin & Mardsen [4] and Landau & Lifshitz [17], more extensively in Batchelor [2], and in the most concrete and elementary way in Feynman volume II [13]. A related topic, the Bernoulli equation, is discussed in Ockendon & Ockendon [20]. This is in essence a first integral of the Euler PDE for irrotational flows.

Batchelor [2] contains extensive discussions of kinematics of flows. Of particular relevance to the problems in this text are the discussion of incompressible flows leading to the stream function in two dimensions and the determination of the velocity field as a functional of the vorticity. Vorticity, vortex dynamics, and related topics such as the circulation theorem (Problems 6.4 and 6.7–6.9 in this text) are prominent in Chorin & Mardsen [4], Feynman volume II [13], and Saffman [21]. In particular, Saffman [21] presents his original work on elliptical vortex patches subject to strain.

Jump conditions across a shock wave are discussed in Chorin & Mardsen [4], Landau & Lifshitz [17], Lighthill [18], Ockendon & Ockendon [20], and Whitham [24]. In particular, the result of Problem 6.4 in this text (that heating induced by a weak shock scales with the cube of the pressure jump) is discussed in Lighthill [18]. Discussion of sound propagation in ideal fluid appears in Chorin & Mardsen [4] (briefly), Ockendon & Ockendon [20], and Whitham [24]. Analyses of wavefronts (surfaces with discontinuous derivatives) are discussed in Courant & Hilbert volume II [9] and Whitham

[**24**]. Problem 6.6 (on linearized supersonic flow about a wing) is motivated by examples in Landau & Lifshitz [**17**].

Finally, the examples on scaling-based reduction (the incompressible limit of compressible flow and Problem 6.10 on buoyancy force) are of my own devising. They are similar in technique to J. Stoker's derivations of shallow water theory from the full free boundary problem, referenced in Chapter 7. The analysis of stationary internal waves in Problem 6.11, based on the results of Problem 6.10, is also of my own devising, but its physical content is very similar to the problem of free surface waves produced by an obstacle. That problem is discussed in Lighthill [**18**].

Free surface waves

The nominal subject of this chapter is the surface waves on oceans and rivers. Their phenomena are close to "everyday experience" and very rich. Their intrinsic interest granted, we come to the real purpose: of "descending" from a fully inclusive model down through a sequence of scaling-based reductions. The reduced models are fundamental sources of intuition in continuum mechanics.

Basic equations. Incompressible fluid of uniform density ρ fills an ocean. In cartesian coordinates (x_1, x_2, x_3), $x_3 = 0$ represents a solid bottom. The upper free surface is represented by

$$(7.1) \qquad x_3 = h(x_1, x_2, t).$$

The velocity field $\boldsymbol{u}(\boldsymbol{x}, t)$ and pressure field $p(\boldsymbol{x}, t)$ in $0 < x_3 < h(x_1, x_2, t)$ satisfy incompressible flow equations

$$(7.2) \qquad \boldsymbol{\nabla} \cdot \boldsymbol{u} = 0,$$

$$(7.3) \qquad \boldsymbol{u}_t + (\boldsymbol{u} \cdot \boldsymbol{\nabla})\boldsymbol{u} = -\frac{1}{\rho}\nabla p - g\boldsymbol{e}_3.$$

In (7.3), $-g\boldsymbol{e}_3$ is the force per unit mass due to gravity. There are boundary conditions. The flow is tangential on the bottom, so

$$(7.4) \qquad u_3 = 0 \quad \text{on } x_3 = 0.$$

On the free surface there is a so-called *kinematic* boundary condition: Let $\boldsymbol{x}(t)$ be a material point on the surface. The continuity of the flow map guarantees that material points initially on the surface remain on the surface

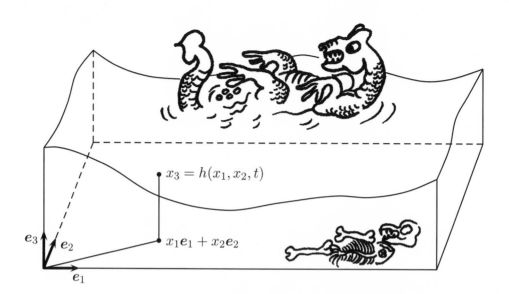

Figure 7.1.

for all time. Hence,

$$(7.5) \qquad\qquad x_3(t) = h(x_1(t), x_2(t), t)$$

for all t. Differentiating (7.5) with respect to t and setting $\dot{x}_i(t) = u_i(\boldsymbol{x}(t), t)$ gives a relation

$$(7.6) \qquad\qquad u_3 = h_t + u_1 \partial_1 h + u_2 \partial_2 h,$$

which holds at each $\boldsymbol{x}(t)$ on the surface. Hence, (7.6) holds on $x_3 = h(x_1, x_2, t)$ for all t, and this is the kinematic boundary condition. There is also a *dynamical* boundary condition. Assume that the medium above the ocean has uniform pressure, which without loss of generality we can take to be zero. If the pressure just below the surface is nonzero, an infinitesimal layer of fluid experiences a finite force per unit area and we get an infinite acceleration of the surface. This doesn't happen. Hence, the pressure field satisfies the dynamical boundary condition

$$(7.7) \qquad\qquad p(x_1, x_2, h(x_1, x_2, t), t) = 0$$

for all x_1, x_2 and t. The PDEs (7.2) and (7.3) together with the boundary conditions (7.4), (7.6) and (7.7) govern the evolution of the surface elevation h, velocity $\boldsymbol{u}$, and pressure p in $0 < x_3 < h(x_1, x_2, t)$.

It is convenient to distinguish the horizontal and vertical dimensions of space in these equations. Write the position vector in $\mathbb{R}^3$ as $\boldsymbol{x} + x_3 \boldsymbol{e}_3$, where $\boldsymbol{x}$ now denotes horizontal displacement, $\boldsymbol{x} := x_1 \boldsymbol{e}_1 + x_2 \boldsymbol{e}_2$. Similarly, write the velocity field as $\boldsymbol{u} + u_3 \boldsymbol{e}_3$, where $\boldsymbol{u} := u_1 \boldsymbol{e}_1 + u_2 \boldsymbol{e}_2$. The gradient operator is represented as $\boldsymbol{\nabla} + \partial_3 \boldsymbol{e}_3$, where $\boldsymbol{\nabla} := \partial_1 \boldsymbol{e}_1 + \partial_2 \boldsymbol{e}_2$. The PDEs (7.2) and (7.3) are reformulated as

$$(7.8) \qquad \boldsymbol{\nabla} \cdot \boldsymbol{u} + \partial_3 u_3 = 0,$$

$$(7.9) \qquad \boldsymbol{u}_t + (\boldsymbol{u} \cdot \boldsymbol{\nabla} + u_3 \partial_3) \boldsymbol{u} = -\frac{1}{\rho} \boldsymbol{\nabla} p,$$

$$(7.10) \qquad \partial_t u_3 + (\boldsymbol{u} \cdot \boldsymbol{\nabla} + u_3 \partial_3) u_3 = -\frac{1}{\rho} \partial_3 p - g$$

in $0 < x_3 < h(\boldsymbol{x}, t)$, and the boundary conditions (7.4), (7.6) and (7.7) become

$$(7.11) \qquad u_3 = 0$$

on $x_3 = 0$ and

$$(7.12) \qquad h_t + \boldsymbol{u} \cdot \boldsymbol{\nabla} h = u_3,$$

$$(7.13) \qquad p = 0$$

on $x_3 = h(\boldsymbol{x}, t)$.

Linearized waves. The simplest solutions of (7.8)–(7.13) are rest states with $h \equiv H$, a uniform constant, $\boldsymbol{u} \equiv 0$, and $p = \rho g(H - x_3)$. The physics of the pressure field is simple. The pressure on the surface $x_3 = H$ is zero, and the pressure at x_3, $0 < x_3 < H$, is what is required to support the weight of the water above the given x_3. *Perturbations* from the rest state are represented by

$$(7.14) \qquad \begin{aligned} \varepsilon h' &:= h - H, \\ \varepsilon \boldsymbol{u}' &:= \boldsymbol{u}, \quad \varepsilon u_3' := u_3, \\ \varepsilon p' &:= p - \rho g(H - x_3). \end{aligned}$$

Here $\varepsilon > 0$ is a gauge parameter, and we consider the limit $\varepsilon \to 0$ with the primed variables fixed (so that perturbations of the dependent variables from their rest values are $O(\varepsilon)$). Writing the PDEs (7.8)–(7.10) in terms of primed variables and taking the limit $\varepsilon \to 0$ gives

$$(7.15) \qquad \begin{aligned} \boldsymbol{\nabla} \cdot \boldsymbol{u}' + \partial_3 u_3' &= 0, \\ \boldsymbol{u}_t' &= -\frac{1}{\rho} \boldsymbol{\nabla} p', \\ \partial_t u_3 &= -\frac{1}{\rho} \partial_3 p' \end{aligned}$$

in the $\varepsilon \to 0$ *limit region*, $0 < x_3 < H$. The boundary condition on the bottom is $u_3 = 0$ on $x_3 = 0$ as in (7.11). The $\varepsilon \to 0$ limits of the surface

boundary conditions (7.12) and (7.13) require some care. The kinematic
boundary condition (7.12) in terms of primed variables is

$$(\varepsilon h'_t + \varepsilon^2 \boldsymbol{u}' \cdot \boldsymbol{\nabla} h')(\boldsymbol{x}, H + \varepsilon h', t) = \varepsilon u_3(\boldsymbol{x}, H + \varepsilon h', t).$$

Taking the limit $\varepsilon \to 0$ gives

(7.16) $h'_t = u'_3 \quad \text{on } x_3 = H.$

The dynamical boundary condition (7.13) translates into

$$\rho g(H - (H + \varepsilon h'(\boldsymbol{x}, H + \varepsilon h', t))) + \varepsilon p'(\boldsymbol{x}, H + \varepsilon h', t) = 0,$$

and the $\varepsilon \to 0$ limit is

(7.17) $p' = \rho g h' \quad \text{on } x_3 = H.$

Notice that the linearized free surface boundary conditions apply on the
unperturbed free surface $x_3 = H$. In particular, notice that p' is *not* zero on
$x_3 = H$. If $h' > 0$, then $\varepsilon p' = \varepsilon \rho g h'$ represents the pressure field on $x_3 = H$
due to the weight of the water in $0 < x_3 < \varepsilon h'$. If $h' < 0$, then $x_3 = H$ is
above the free surface, and $\varepsilon p' = \varepsilon \rho g h'$ represents a linear extrapolation of
the pressure field to $x_3 = H$.

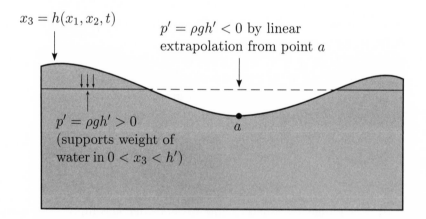

Figure 7.2.

Plane waves and the dispersion relation. The linearized equations
(7.15)–(7.17) have plane wave solutions with

(7.18)
$$h' = h^0 e^{i(\boldsymbol{k} \cdot \boldsymbol{x} - \omega t)},$$
$$\boldsymbol{u}' = \boldsymbol{u}^0(x_3) e^{i(\boldsymbol{k} \cdot \boldsymbol{x} - \omega t)},$$
$$u'_3 = u^0_3(x_3) e^{\boldsymbol{k} \cdot \boldsymbol{x} - \omega t},$$
$$p' = p^0(x_3) e^{i(\boldsymbol{k} \cdot \boldsymbol{x} - \omega t)}.$$

Since $\partial_t e^{i(\boldsymbol{k}\cdot\boldsymbol{x}-\omega t)} = -i\omega e^{i(\boldsymbol{k}\cdot\boldsymbol{x}-\omega t)}$ and $\boldsymbol{\nabla} e^{i(\boldsymbol{k}\cdot\boldsymbol{x}-\omega t)} = i\boldsymbol{k}e^{i(\boldsymbol{k}\cdot\boldsymbol{x}-\omega t)}$, we obtain reduced equations for h^0, $\boldsymbol{u}^0(x_3)$, $u_3^0(x_3)$ and $p^0(x_3)$ via the replacements $\boldsymbol{\nabla} \to i\boldsymbol{k}$ and $\partial_t \to -i\omega$ in the linearized equations (7.15)–(7.17). The reduced equations are

$$i\boldsymbol{k} \cdot \boldsymbol{u}^0 + \partial_3 u_3^0 = 0,$$

(7.19)
$$-i\omega\boldsymbol{u}^0 = -\frac{1}{\rho}i\boldsymbol{k}p^0,$$

(7.20)
$$-i\omega u_3^0 = -\frac{1}{\rho}\partial_3 p^0$$

in $0 < x_3 < H$, with boundary conditions

$$u_3^0 = 0$$

on $x_3 = 0$ and

(7.21)
$$u_3^0 = -i\omega h^0,$$
$$p^0 = \rho g h^0$$

on $x_3 = H$. Assuming $\omega \neq 0$, we can eliminate $\boldsymbol{u}^0$ and u_3^0 to extract a boundary value problem for $p^0(x_3)$ which contains h^0 as a parameter:

$$\partial_{33}p^0 - (\boldsymbol{k} \cdot \boldsymbol{k})p^0 = 0 \quad \text{in } 0 < x_3 < H,$$
$$\partial_3 p^0 = 0 \quad \text{on } x_3 = 0,$$
$$p^0 = \rho g h^0 \quad \text{on } x_3 = H.$$

The solution for $p^0(x_3)$ is

(7.22)
$$p^0(x_3) = \rho g h_0 \frac{\cosh kx_3}{\cosh kH}.$$

Here, $k := \sqrt{\boldsymbol{k} \cdot \boldsymbol{k}}$. It follows from (7.20) and (7.21) that

$$\frac{1}{\rho}\partial_3 p^0 = \omega^2 h^0$$

on $x_3 = H$, and upon substituting (7.22) for $p^0(x_3)$, we get the *dispersion relation*

(7.23)
$$\omega^2 = gk \tanh kH,$$

which gives two "branches" of ω as functions of k. For the branch with $\omega > 0$, the phase speed $v := \frac{\omega}{k}$ as a function of k is

(7.24)
$$v = \sqrt{gH\frac{\tanh kH}{kH}}.$$

Figure 7.3 is a graph of the phase speed as a function of k. In the *long wave limit* $kH \to 0$, v asymptotes to the uniform constant $\sqrt{gH}$. In the deep water limit $kH \to \infty$, v asymptotes to $\sqrt{g/k}$, represented by the dashed line in Figure 7.3.

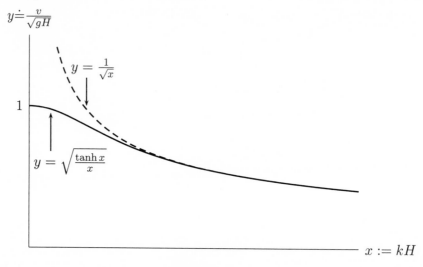

Figure 7.3.

We look at the particle kinematics of linearized plane waves. Assume that the wave propagates in the e_1 direction, with $k = ke_1$. From (7.19), (7.20) and (7.22) we find that the $\mathbb{R}^3$ velocity field is

$$
\begin{aligned}
u' &= \mathrm{Re}\{(u^0 + u_3^0 e_3)e^{i(k \cdot x - \omega t)}\} \\
&= \frac{1}{\omega \rho} \mathrm{Re}\{(kp^0 e_1 - i\partial_3 p^0 e_3)e^{i(k \cdot x - \omega t)}\} \\
&= \frac{g}{\omega} \frac{h^0}{H} \frac{kH}{\cosh kH}\{\cosh kx_3 \cos(kx_1 - \omega t)e_1 \\
&\qquad\qquad + \sinh kx_3 \sin(kx_1 - \omega t)e_3\}.
\end{aligned}
$$

(7.25)

This velocity field translates with the wave. Figure 7.4a is a visualization of u at a fixed moment in time.

Particle trajectories $x(t)$ are solutions of the ODE $\dot{x}(t) = \varepsilon u'(x(t), t)$. In the limit $\varepsilon \to 0$, with t fixed, the variation $\delta x(t)$ of $x(t)$ is $O(\varepsilon)$. Hence, in the limit $\varepsilon \to 0$, an approximation to the variation is determined by integrating $\varepsilon u'(x, t)$ in t with x fixed. Thus,

$$\delta x(t) = \frac{g}{\omega^2} \frac{\varepsilon h^0}{H} \frac{kH}{\cosh kH}\{\cosh kx_3 \sin(\omega t - kx_1)e_1 + \sinh kx_3 \cos(\omega t - kx_1)e_3\}.$$

The prefactor can be simplified by substituting for ω^2 from the dispersion relation (7.23). Upon doing this we obtain
(7.26)

$$\delta x(t) = \varepsilon h^0 \left\{ \frac{\cosh kx_3}{\sinh kH} \sin(\omega t - kx_1)e_1 + \frac{\sinh kx_3}{\sinh kH} \cos(\omega t - kx_1)e_3 \right\}.$$

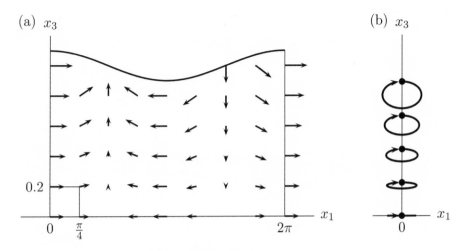

Figure 7.4.

Note that at the surface $(x_3 = H)$, the vertical component of displacement reduces to $\varepsilon h^0 \cos(\omega t - kx_1)$, as it should. The trajectories described by (7.26) are clockwise ellipses. Figure 7.4b depicts the ellipses corresponding to particles which started from the x_3-axis at $t = 0$. With a little imagination, you can see that these clockwise ellipses are consistent with the velocity field in Figure 7.4a translating to the right.

Problem 7.1 (Internal waves, again).   In Problem 6.11 we analyzed stationary waves in stably stratified fluid. We eventually specialized to waves on the interface between dense fluid below and light fluid above. The analysis was based on the asymptotic limit of small density difference. Here, we re-examine these stationary interfacial waves as a free boundary problem, but this time with *no* assumption on the smallness of the density difference.

a)   Formulate the free boundary problem for two-dimensional, time-independent, incompressible ideal fluid flow, with density $\rho = \rho_-$, a positive constant, in $0 < x_2 < h(x_1)$, and density $\rho = \rho_+$ in $x_2 > h(x_1)$, where ρ_+ is another positive constant with $\rho_+ < \rho_-$. There is a gravity force of $-ge_2$ per unit mass. The plane $x_2 = 0$ is a solid bottom.

   Linearize about the uniform flow solution $\boldsymbol{u} \equiv U\boldsymbol{e}_1$ in $x_2 > 0$ and $h(x_1) \equiv H$, a positive constant. Derive a "reduced" solution of the linearized free boundary problem containing only the interface and pressure perturbations.

b)   Analyze the reduced free boundary problem in Part a for stationary wave solutions whose x_1-dependence is the exponential e^{ikx_1}. Determine the relation between the unperturbed flow velocity U and the wavenumber k. Compare this to the result in Problem 6.11b. Assuming we have U such that k is real, construct approximate streamlines of the flow and graph them in the (x_1, x_2) plane.

Solution.

a) The PDEs in $0 < x_2 < h(x_1)$ and in $x_2 > h(x_1)$ are

$$\boldsymbol{\nabla} \cdot \boldsymbol{u} = 0,$$

$$(\boldsymbol{u} \cdot \boldsymbol{\nabla})\boldsymbol{u} = -g\boldsymbol{e}_2 - \frac{1}{\rho}\boldsymbol{\nabla}p,$$

where $\rho \equiv \rho_-$ in $0 < x_2 < h(x_1)$ and $\rho \equiv \rho_+$ in $x_2 > h(x_1)$. The "bottom" boundary condition is

$$u_2 = 0$$

on $x_2 = 0$. The kinematic boundary conditions on the interface are

$$u_2(x_1, h(x_1)^+) = h'(x_1)u_1(x_1, h(x_1)^+),$$
$$u_2(x_1, h(x_1)^-) = h'(x_1)u_1(x_1, h(x_1)^-).$$

In addition, the pressure is continuous across the interface, so

$$p(x_1, h(x_1)^+) = p(x_1, h(x_1)^-).$$

The pressure field associated with the uniform flow solution is

$$p = p(x_2) := \left[\begin{array}{ll} -g\rho_-(x_2 - H), & 0 < x_2 < H, \\ -g\rho_+(x_2 - H), & x_2 > H. \end{array} \right.$$

Next, define perturbation quantities

$$\varepsilon h' := h - H,$$
$$\varepsilon \boldsymbol{u}' := \boldsymbol{u} - U\boldsymbol{e}_1,$$
$$\varepsilon p' := p - p(x_2).$$

Here, ε is a positive gauge parameter, and we will consider the limit $\varepsilon \to 0$. Writing the PDE in terms of perturbation variables and taking $\varepsilon \to 0$ gives

$$\partial_1 u_1' + \partial_2 u_2' = 0',$$

(7.1-1) $$U\partial_1 u_1' = -\frac{1}{\rho}\partial_1 p',$$

$$U\partial_1 u_2' = -\frac{1}{\rho}\partial_2 p'.$$

Since the interface perturbation from $h = H$ vanishes as $\varepsilon \to 0$, the density ρ in (7.1-1) is ρ_- in $0 < x_2 < H$ and ρ_+ in $x_2 > H$. The bottom boundary condition translates into

$$(7.1\text{-}2) \qquad u'_2(x_1, 0) = 0,$$

and the kinematic boundary conditions in terms of perturbation variables reduce to

$$(7.1\text{-}3) \qquad u_2(x_1, H^+) = u_2(x_1, H^-) = U\partial_1 h'(x)$$

in the limit $\varepsilon \to 0$. The continuity of pressure across the interface is expressed exactly by

$$-\varepsilon g \rho_- h'(x_1) + \varepsilon p'(x_1, (H + \varepsilon h'(x_1))^-) = -\varepsilon g \rho_+ h'(x_1) + \varepsilon p'(x_1, (H + \varepsilon h'(x_1))^+).$$

In the limit $\varepsilon \to 0$ this reduces to

$$(7.1\text{-}4) \qquad p'(x_1, H^+) - p'(x_1, H^-) = g(\rho_+ - \rho_-)h'(x_1).$$

The full set of linearized equations consists of the PDEs (7.1-1) and the boundary conditions (7.1-2)–(7.1-4).

We derive reduced equations for the pressure and interface perturbations p' and h': First, from (7.1-1) it follows that p' is harmonic, i.e.,

$$(7.1\text{-}5) \qquad \Delta p' = 0$$

in $0 < x_2 < H$ and in $x_2 > H$. From the bottom boundary condition (7.1-2) and the e_2-component of the linearized Euler equation (the last equation of (7.1-1)), we find that

$$(7.1\text{-}6) \qquad \partial_2 p'(x, 0) = 0.$$

From the kinematic boundary conditions (7.1-3) and the e_2-component of the linearized Euler equation, we further have

$$(7.1\text{-}7) \qquad U^2 \partial_{11} h(x_1) = -\frac{1}{\rho_-} \partial_2 p'(x_1, H^-) = -\frac{1}{\rho_+} \partial_2 p'(x_1, H^+).$$

In summary, the harmonic pressure perturbation is subject to the "bottom" boundary condition (7.1-6) and *three* interface conditions: the two in (7.1-7), and (7.1-4). We will soon see that we have exactly the right number of boundary conditions.

b) We seek linearized stationary waves with

$$(7.1\text{-}8) \qquad h'(x_1) = h_0 e^{ikx_1},$$

$$(7.1\text{-}9) \qquad p'(x_1, x_2) = p_0(x_2) e^{ikx_1}.$$

Substituting (7.1-9) into Laplace's equation gives an ODE for $p_0(x_2)$,

$$p''_0(x_2) - k^2 p_0(x_2) = 0$$

in $0 < x_2 < H$ and in $x_2 > H$. Solutions which satisfy the bottom boundary
condition (7.1-6) and don't blow up as $x_2 \to +\infty$ are

$$
(7.1\text{-}10) \qquad p_0(x_2) = \left[\begin{array}{ll} p_- \dfrac{\cosh kx_2}{\cosh kH}, & 0 < x_2 < H, \\[3mm] p_+ e^{-k(x_2-H)}, & x_2 > H, \end{array} \right.
$$

where p_-, p_+ are constants. Notice "details of craftsmanship" such as dividing by $\cosh kH$ in $0 < x_2 < H$ and use of the argument $x_2 - H$ in
$x_2 > H$. Substituting (7.1-10) into the three interface conditions in (7.1-4)
and (7.1-7) gives three homogeneous linear equations for the constants h_0,
p_+, p_-:

$$
\begin{aligned}
p_1 - p_+ &= g(\rho_- - \rho_+)h_0, \\[2mm]
kU^2 h_0 &= \frac{p_-}{\rho_-}\tanh kH, \\[2mm]
kU h_0 &= -\frac{p_+}{\rho_+}.
\end{aligned}
$$

$(7.1\text{-}11)$

The consistency condition is

$$
(7.1\text{-}12) \qquad kH\frac{\rho_-\cosh kH + \rho_+\sinh kH}{\bar{\rho}\sinh kH} = \frac{\rho_- - \rho_+}{\bar{\rho}}\frac{gH}{U^2},
$$

where $\bar{\rho} := \frac{\rho_+ + \rho_-}{2}$. This is the relation between the unperturbed flow speed
U and the wavenumber k.

In the limit $\rho_+, \rho_- \to \rho_0$, it is easy to see that (7.1-12) reduces to its
counterpart (6.11-10) from our first crack at internal waves. The qualitative
features of the k versus U relation are very little changed. In particular, kH
decreases to zero ("long waves") as U approaches the speed limit

$$
U_{\max} := \sqrt{\frac{\rho_- - \rho_+}{\rho_-}}\sqrt{gH}
$$

from below.

Streamlines $x_2 = x_2(x_1)$ are integral curves of the velocity field, and as
such, they satisfy the ODE

$$
(7.1\text{-}13) \qquad \frac{dx_2}{dx_1} = \frac{\varepsilon u_2'}{U + \varepsilon u_1'} \sim \frac{\varepsilon}{U}u_2'
$$

as $\varepsilon \to 0$. The u_2' in the right-hand side can be related to p', which is
now known through (7.1-9) and (7.1-10). In the third equation of (7.1-1)
(e_2-component of the linearized Euler equation), set $p' = p_0(x_2)e^{ikx_1}$ and
integrate with respect to x_1 to get

$$
u_2' = -\frac{p_0'(x_2)}{\rho U}\frac{e^{ikx_1}}{ki}.
$$

Substituting the *real* part of u'_2 into (7.1-13) gives

(7.1-14)
$$\frac{dx_2}{dx_1} \sim -\varepsilon \frac{p'_0(x_2)}{\rho U^2} \frac{\sin kx_1}{k},$$

with $p_0(x_2)$ as in (7.1-10). The asymptotic solutions of this ODE as $\varepsilon \to 0$ are

(7.1-15)
$$x_2(x_1) \sim \bar{x}_2 + \varepsilon \frac{p'_0(\bar{x}_2)}{\rho U^2} \frac{\cos kx_1}{k^2},$$

where $\bar{x}_2$ is a constant. In $0 < \bar{x}_2 < H$, (7.1-15) reduces to

$$x_2(x_1) \sim \bar{x}_2 + \varepsilon \frac{p_-}{\rho_- U^2} \frac{\sinh k\bar{x}_2}{k \cosh kH} \cos kx_1.$$

By the second equation of (7.1-11) we can simplify the prefactor of the second term on the right-hand side, so $x_2(x_1) \sim \bar{x}_2 + \varepsilon h_0 \cos kx_1$. Similarly,

$$x_2(x_1) \sim \bar{x}_2 + \varepsilon h_0 e^{-k(x_2-H)} \cos kx_1$$

in $\bar{x}_2 > H$. Figure 7.5 is the graph of the streamlines.

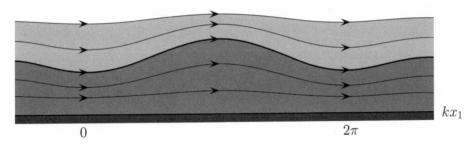

kx_1

0 2π

Figure 7.5.

Shallow water equations. In the long wave limit $kH \to 0$, the phase speed of the plane wave reduces to $\sqrt{gH}$, as noted before, and the velocity field (7.25) reduces to

$$\boldsymbol{u}' = \frac{g}{\omega} k h^0 \cos(kx_1 - \omega t)\boldsymbol{e}_1.$$

This velocity field is horizontal and independent of x_3. This feature of linearized long waves suggests a limit process that reduces the full free boundary problem to *shallow water* PDEs describing long but fully nonlinear waves.

Let H denote the depth of the ocean at rest, as before, and let L be the characteristic horizontal scale of the surface waves. The gauge parameter is $\varepsilon := \frac{H}{L}$ and the long wave limit is $\varepsilon \to 0$. In shallow water theory the characteristic variation of depth is not small compared to H as in the linearized theory, but has the same magnitude as H itself. Hence, the unit of horizontal displacement is L, and the unit of vertical displacement h is $H = \varepsilon L$, as in the scaling table below:

(7.27)

Variable	$\boldsymbol{x}$	x_3, h	t	p	$\boldsymbol{u}$	u_3
Unit	L	εL	$\sqrt{\frac{L}{\varepsilon g}}$	$\varepsilon \rho g L$	$\sqrt{\varepsilon g L}$	$\varepsilon \sqrt{\varepsilon g L}$

The unit of time is $[t] := \frac{L}{\sqrt{gH}} = \sqrt{\frac{L}{\varepsilon g}}$, the time it takes a linearized wave to travel distance L. Of course, nonlinear waves don't necessarily travel at the linearized speed, but the linearized wave speed does give the correct order of magnitude.

It remains to determine the units of the intensive variables p, $\boldsymbol{u}$ and u_3. The hydrostatic pressure at rest, $p = \rho g(H - x_3)$, indicates the correct order of magnitude for pressure, so the unit of pressure is $\rho g H = \varepsilon \rho g L$. The unit $[\boldsymbol{u}]$ of horizontal velocity follows by an order-of-magnitude balance between the horizontal acceleration of fluid particles and the horizontal pressure force per unit mass. That is,

$$\frac{[\boldsymbol{u}]}{[t]} = \frac{[\boldsymbol{u}]}{\sqrt{L/\varepsilon g}} = \frac{1}{\rho L}(\varepsilon \rho g L),$$

so $[\boldsymbol{u}] = \sqrt{\varepsilon g L}$, the same as the long wave speed $\sqrt{gH}$. Finally, $[u_3]$ should be the characteristic variation of depth divided by the unit of time, so

$$[u_3] = \frac{H}{[t]} = \frac{\varepsilon L}{\sqrt{L/\varepsilon g}} = \varepsilon \sqrt{\varepsilon g L}.$$

The above units of p, $\boldsymbol{u}$ and u_3 are entered in the scaling table (7.27).

The dimensionless versions of the basic equations (7.8)–(7.13) are

(7.28)
$$\boldsymbol{\nabla} \cdot \boldsymbol{u} + \partial_3 u_3 = 0,$$

(7.29)
$$\partial_t \boldsymbol{u} + (\boldsymbol{u} \cdot \boldsymbol{\nabla} + u_3 \partial_3)\boldsymbol{u} = -\boldsymbol{\nabla} p,$$

(7.30)
$$\varepsilon^2 \{ \partial_t u_3 + (\boldsymbol{u} \cdot \boldsymbol{\nabla} + u_3 \partial_3) u_3 \} = -\partial_3 p - 1$$

in $0 < x_3 < h(\boldsymbol{x}, t)$,

(7.31)
$$u_3 = 0$$

on $x_3 = 0$, and

(7.32)
$$p = 0,$$

(7.33)
$$u_3 = \partial_t h + (\boldsymbol{u} \cdot \boldsymbol{\nabla})h$$

on $x_3 = h$. In the limit $\varepsilon \to 0$, only the vertical momentum equation (7.30) has an obvious reduction to

$$(7.34) \qquad\qquad\qquad \partial_3 p = -1.$$

The remaining equations formally retain all their terms. But the stage is set for massive reduction. It follows from (7.34) and $p = 0$ on $x_3 = h$ that

$$(7.35) \qquad\qquad\qquad p = h(\boldsymbol{x}, t) - x_3.$$

Hence, the hydrostatic approximation to pressure applies asymptotically in the shallow water dynamics. Substituting (7.35) into the horizontal momentum PDE (7.29) gives

$$(7.36) \qquad\qquad \partial_t \boldsymbol{u} + (\boldsymbol{u} \cdot \boldsymbol{\nabla} + u_3 \partial_3)\boldsymbol{u} = -\boldsymbol{\nabla} h.$$

The right-hand side $\boldsymbol{\nabla} h$ is independent of x_3. Hence, the horizontal acceleration of fluid particles is independent of x_3. Suppose a region of ocean is at rest at time $t = 0$, before waves arrive. Then fluid particles in any vertical line remain in a vertical line for all $t > 0$. This happens only if the horizontal velocity $\boldsymbol{u}$ is independent of x_3, i.e., $\boldsymbol{u} = \boldsymbol{u}(\boldsymbol{x}, t)$. More generally, if $\boldsymbol{u}$ is initially independent of x_3, it remains independent of x_3 for all time. Under these conditions, (7.36) reduces to

$$(7.37) \qquad\qquad\qquad \partial_t \boldsymbol{u} + (\boldsymbol{u} \cdot \boldsymbol{\nabla})\boldsymbol{u} = -\boldsymbol{\nabla} h.$$

Given $\boldsymbol{u}$ independent of x_3, it now follows from the incompressibility equation (7.28) and the boundary condition $u_3 = 0$ on $x_3 = 0$ that

$$u_3(\boldsymbol{x}, h, t) = -x_3 \boldsymbol{\nabla} \cdot \boldsymbol{u},$$

and substituting this result into the kinematic boundary condition (7.33) gives

$$-h\boldsymbol{\nabla} \cdot \boldsymbol{u} = \partial_t h + (\boldsymbol{u} \cdot \boldsymbol{\nabla})h,$$

or

$$(7.38) \qquad\qquad\qquad h_t + \boldsymbol{\nabla} \cdot (h\boldsymbol{u}) = 0.$$

We recognize (7.37) and (7.38) as a pair of PDEs for the ocean's elevation $h(\boldsymbol{x}, t)$ and horizontal velocity field $\boldsymbol{u}(\boldsymbol{x}, t)$. These are the famous *shallow water equations*. Notice that they are formally equivalent to the two-dimensional ideal fluid PDEs (6.18) and (6.19) with h in the role of density and $\frac{1}{2}h^2$ in the role of pressure.

Problem 7.2 (Variable-depth ocean and "slow currents").   We consider an ocean whose depth at rest is a function of horizontal position $\boldsymbol{x}$. The characteristic length of $\boldsymbol{x}$-variations in the bottom topography is L, and the characteristic depth is $H \ll L$, so we have the small gauge parameter $\varepsilon := \frac{H}{L}$ (the same as before). The bottom is given by

$$(7.2\text{-}1) \qquad\qquad \frac{x_3}{H} = -\ell\left(\frac{\boldsymbol{x}}{L}\right).$$

a) Derive modified shallow water equations which include the effect of the variable bottom topography.

b) We propose a simplified model of "slow currents" in our variable-depth ocean. The characteristic horizontal scale of the currents is L, the same as that of the bottom topography. The scaling unit $[u]$ of horizontal velocity is much smaller than the characteristic speed $\sqrt{gH}$ of surface waves. The characteristic time is $L/[u]$.

   We could perform a "slow current" scaling reduction, starting from the full dimensional equations. But it is simpler to do a "secondary scaling" of the variable-depth shallow water equations from Part a. The formal limit process begins by assuming that the dimensionless horizontal velocity $\boldsymbol{u}$ has order of magnitude μ, where μ is a small gauge parameter (independent of $\varepsilon := \frac{H}{L}$). The brief physical description of the "slow current" limit in the preceding paragraph indicates what the collateral scaling of $\boldsymbol{x}$, t and h with μ are. Derive asymptotic $\mu \to 0$ equations of "slow current flow".

Solution.

a)   The dimensionless PDEs for $\boldsymbol{u}$, u_3 and p are (7.28)–(7.30), the same as before, but now they apply in $-\ell(\boldsymbol{x}) < x_3 < h(\boldsymbol{x}, t)$. The free surface conditions (7.32) and (7.33) also apply with no change. The only change is the replacement of the bottom boundary condition (7.31) by

$$(7.2\text{-}2) \qquad\qquad u_3 = -(\boldsymbol{u} \cdot \boldsymbol{\nabla})\ell$$

on $x_3 = -\ell(\boldsymbol{x})$. The argument leading to the "shallow water Euler equation" (7.37) is completely unchanged, so

$$(7.2\text{-}3) \qquad\qquad \boldsymbol{u}_t + (\boldsymbol{u} \cdot \boldsymbol{\nabla})\boldsymbol{u} = -\boldsymbol{\nabla}h$$

as before. Where, then, does $\ell(\boldsymbol{x})$ appear? Given the new bottom boundary condition (7.2-2), x_3-integration of $\partial_3 u_3 = -\boldsymbol{\nabla} \cdot \boldsymbol{u}$ (from (7.28)) gives

$$u_3 = -(\boldsymbol{u} \cdot \boldsymbol{\nabla})\ell - (\boldsymbol{\nabla} \cdot \boldsymbol{u})(x_3 + \ell),$$

and substituting this result for u_3 into the kinematic boundary condition (7.33) on the free surface, we get

$$h_t + (\boldsymbol{u} \cdot \boldsymbol{\nabla})h = -(\boldsymbol{u} \cdot \boldsymbol{\nabla})\ell - (\boldsymbol{\nabla} \cdot \boldsymbol{u})(h + \ell),$$

which consolidates into

(7.2-4) $$h_t + \boldsymbol{\nabla} \cdot \{(\ell + h)\boldsymbol{u}\} = 0.$$

The modified shallow water equations are (7.2-3) and (7.2-4). The only change is in replacing the volume flux $h\boldsymbol{u}$ in the volume conservation PDE (7.38) by $(\ell + h)\boldsymbol{u}$.

b) We are given μ as the scaling unit of $\boldsymbol{u}$, i.e., $[u] = \mu$. Since the characteristic horizontal scale of the currents is the same as that of the bottom topography, there is *no* scaling of $\boldsymbol{x}$ with μ. The characteristic time to make a unit horizontal displacement at speed $[u] = \mu$ is $[t] = \frac{1}{\mu}$. The scaling unit $[h]$ of h is determined by dominant balance in the rescaled version of the "shallow water Euler equation" (7.2-3). Following the nondimensionalization procedure, we have

$$\frac{[u]}{[t]}\boldsymbol{u}_t + [u]^2(\boldsymbol{u} \cdot \boldsymbol{\nabla})\boldsymbol{u} = -[h]\boldsymbol{\nabla}h$$

or, using $[u] = \mu$ and $[t] = \frac{1}{\mu}$,

$$\mu^2\{\boldsymbol{u}_t + (\boldsymbol{u} \cdot \boldsymbol{\nabla})\boldsymbol{u}\} = -[h]\boldsymbol{\nabla}h;$$

so the scaling unit of h is $[h] = \mu^2$, and thus (7.2-3) is *invariant* under our slow current rescaling. It is the volume conservation PDE (7.2-4) which simplifies. Applying nondimensionalization to (7.2-4) gives

$$\frac{[h]}{[t]}h_t + [u]\boldsymbol{\nabla} \cdot \{(\ell + [h]h)\boldsymbol{u}\} = 0,$$

or

$$\mu^2 h_t + \boldsymbol{\nabla} \cdot \{(\ell + \mu^2 h)\boldsymbol{u}\} = 0,$$

and in the limit $\mu \to 0$,

(7.2-5) $$\boldsymbol{\nabla} \cdot (\ell\boldsymbol{u}) = 0.$$

In summary, the slow current equations are (7.2-3) and (7.2-5). The reduced volume conservation PDE (7.2-5) is what we should get when the departure of the free surface from "rest" ($x_3 \equiv 0$) can be neglected.

Problem 7.3 (What bottom topography does to vortices).   We present a "vorticity formulation" of the slow current equations (7.2-3) and (7.2-5).

a) Let $\omega := \partial_1 u_2 - \partial_2 u_1$ be the vorticity associated with the two-dimensional flow $\boldsymbol{u}$. Show that $\frac{\omega}{\ell}$ is a convected scalar.

b)   To determine a "vorticity evolution equation", which specifies ω_t as a functional of ω, it is sufficient to determine the velocity field $\boldsymbol{u}$ as a functional of ω. The key idea is a generalized stream function: We assume that our "ocean" spans a bounded, simply connected region R of $\mathbb{R}^2$. Let $\boldsymbol{x}_0$ be a fixed point on ∂R. Let $\boldsymbol{x}$ be any point in the interior of R, and let C be any non-self-intersecting curve in R connecting $\boldsymbol{x}_0$ to $\boldsymbol{x}$. Explain why the volume of fluid crossing C per unit time depends only on the endpoint $\boldsymbol{x}$, so that it is a well-defined function $\psi = \psi(\boldsymbol{x})$ of $\boldsymbol{x}$. This is our generalized stream function. Explain why ψ vanishes on ∂R. How do you recover $\boldsymbol{u}$ from ψ? Express the vorticity ω in terms of ψ. Identify the boundary value problem whose solution determines ψ as a functional of ω.

Solution.

a)   The "slow current" Euler equation (7.2-3) expresses the acceleration of fluid particles as the gradient of a scalar, $\frac{D\boldsymbol{u}}{Dt} = -\boldsymbol{\nabla} h$. This is the essential ingredient in the proof of the circulation theorem (Problem 6.4), so the circulation theorem applies here: The circulation

$$(7.3\text{-}1) \qquad\qquad P := \int_{C(t)} \boldsymbol{u} \cdot d\boldsymbol{x}$$

about a closed material curve $C(t)$ is time-independent. By Green's theorem, we convert the line integral (7.3-1) into an area integral of vorticity ω over the interior $S(t)$ of $C(t)$, so we have

$$\frac{d}{dt} \int_{S(t)} \omega \, d\boldsymbol{x} = 0,$$

and in the limit of $S(t)$ shrinking to a point, we deduce that

$$(7.3\text{-}2) \qquad\qquad \frac{D\omega}{Dt} = -(\boldsymbol{\nabla} \cdot \boldsymbol{u})\omega.$$

Next, rewrite (7.2-5) as

$$(7.3\text{-}3) \qquad\qquad \boldsymbol{\nabla} \cdot \boldsymbol{u} = -\frac{\boldsymbol{u} \cdot \boldsymbol{\nabla}\ell}{\ell}.$$

Since $\ell = \ell(\boldsymbol{x})$ is independent of t,

$$\frac{D\ell}{Dt} = \boldsymbol{u} \cdot \boldsymbol{\nabla}\ell.$$

Hence, (7.3-3) becomes

$$(7.3\text{-}4) \qquad \nabla \cdot \boldsymbol{u} = -\frac{1}{\ell}\frac{D\ell}{Dt},$$

and substituting (7.3-4) for $\nabla \cdot \boldsymbol{u}$ into (7.3-2), we obtain

$$\frac{1}{\ell}\frac{D\ell}{Dt} = \frac{1}{\omega}\frac{D\omega}{Dt}.$$

It follows that

$$\frac{D}{Dt}\left(\frac{\omega}{\ell}\right) = 0,$$

so $\frac{\omega}{\ell}$ is constant along particle paths.

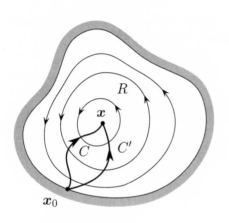

Figure 7.6.

b) Figure 7.6 depicts a portion of R, covered by integral curves of the velocity field $\boldsymbol{u}$, and two curves C, C' from $\boldsymbol{x}_0$ to $\boldsymbol{x}$.

If the rates of fluid volume crossing C and C' were different, volume would accumulate in the interior S of $C - C'$; but within the slow current approximation, the free surface is flat, and the volume of fluid in S is $\int_S \ell(\boldsymbol{x})\, d\boldsymbol{x}$, a fixed constant. If C and C' intersect at other points besides $\boldsymbol{x}_0$ and $\boldsymbol{x}$, we could repeat the argument for each closed "loop" of $C - C'$. So we have our generalized stream function $\psi(\boldsymbol{x})$. Suppose $\boldsymbol{x}$ is on ∂R. Take C to be a curve in R connecting $\boldsymbol{x}_0$ to $\boldsymbol{x}$ and dividing R into two pieces. The volume of fluid in each piece is constant, so the rate of fluid volume crossing C is zero, and hence $\psi(\boldsymbol{x}) = 0$.

The line integral representation of $\psi(\boldsymbol{x})$ is

$$(7.3\text{-}5) \qquad \psi(\boldsymbol{x}) = \int_C \ell \boldsymbol{u} \cdot \boldsymbol{n}\, ds = \int_C -\ell u_2\, dx_1 + \ell u_1\, dx_2,$$

where C is any curve in R going from $\boldsymbol{x}_0$ to $\boldsymbol{x}$. In the right-hand side, we must have $-\ell u_2 = \partial_1 \psi$ and $\ell u_1 = \partial_2 \psi$, so

$$(7.3\text{-}6) \qquad u_1 = \frac{1}{\ell}\partial_2 \psi, \quad u_2 = -\frac{1}{\ell}\partial_1 \psi.$$

From (7.3-6), we further deduce that

$$\omega := \partial_1 u_2 - \partial_2 u_1 = -\partial_1\left(\frac{1}{\ell}\partial_1 \psi\right) - \partial_2\left(\frac{1}{\ell}\partial_2 \psi\right),$$

or finally,

$$(7.3\text{-}7) \qquad\qquad \boldsymbol{\nabla} \cdot \left(\frac{\boldsymbol{\nabla}\psi}{\ell} \right) = -\omega$$

in R. The boundary value problem for ψ consists of the PDE (7.3-7) and the zero boundary condition on ψ along ∂R.

Problem 7.4 (A cautionary tale about point vortices).   This problem deals with a specific example of the vortex dynamics in Problem 7.3.

a) The "ocean" space is $\mathbb{R}^2$, and the depth function is $\ell(\boldsymbol{x}) = e^{-x_1}$. Compute the generalized stream function ψ due to a point vortex of unit circulation at $\boldsymbol{x} = 0$, assuming $\boldsymbol{u} \to 0$ at ∞. (Hint: The analysis of the "smoke plume" in Problem 4.3 should help you get started; but this problem is happening in $\mathbb{R}^2$, not $\mathbb{R}^3$. You might want to recall Problem 4.8, in which we did a "descent" from $\mathbb{R}^3$ to $\mathbb{R}^2$. The result of your analysis is going to be an integral representation of ψ.)

b) Compute the average of the velocity field over a circle of radius ε about $\boldsymbol{x} = 0$, in the limit $\varepsilon \to 0$. (Hint: The analysis in Problem 4.9 suggests the first steps.)

c) What we really want is nearly uniform vorticity inside the disk $r < \varepsilon$ about $\boldsymbol{x} = 0$. We make the assumption that the flow due to this vortex patch in $r < \varepsilon$ is well approximated by the point vortex flow analyzed in Part a. Give a reasonable conjecture for the motion of this vortex patch as induced by the nonuniform bottom. What happens as $\varepsilon \to 0$?

Solution.

a) For $\ell = e^{-x_1}$ and $\omega = \delta(\boldsymbol{x})$, the PDE (7.3-7) takes the specific form

$$(7.4\text{-}1) \qquad\qquad \Delta\psi + \partial_1\psi = -\delta(\boldsymbol{x}).$$

The "smoke plume" in Problem 4.3 suggests a new dependent variable $g(\boldsymbol{x})$, related to $\psi(\boldsymbol{x})$ by

$$(7.4\text{-}2) \qquad\qquad \psi = e^{-\frac{x_1}{2}} g,$$

where $g(\boldsymbol{x})$ satisfies the PDE

$$\Delta g - \frac{g}{4} = -\delta(\boldsymbol{x}).$$

The relevant solution is circularly symmetric and vanishes at ∞. If we were in $\mathbb{R}^3$, the solution would be

$$g = \frac{e^{-\frac{r}{2}}}{4\pi r}.$$

In $\mathbb{R}^2$, we form a superposition of such three-dimensional solutions along a line, like in Problem 4.8. This gives the integral representation of the two-dimensional solution,

$$(7.4\text{-}3) \qquad g = \frac{1}{4\pi} \int_{-\infty}^{\infty} \frac{e^{-\frac{1}{2}\sqrt{r^2+z^2}}}{\sqrt{r^2+z^2}}\, dz = \frac{1}{2\pi}\int_0^{\infty} e^{-\frac{r}{2}\cosh\zeta}\, d\zeta.$$

b) The velocity field components are given by (7.3-6) with $\ell = e^{-x_1}$, so

$$(7.4\text{-}4) \qquad \boldsymbol{u} = e^{x_1}(\partial_2\psi\boldsymbol{e}_1 - \partial_1\psi\boldsymbol{e}_2) = -e^{x_1}J\boldsymbol{\nabla}\psi,$$

where $J := \begin{bmatrix} 0 & -1 \\ 1 & 0 \end{bmatrix}$ represents rotation by $\frac{\pi}{2}$ radians. For $g = g(r)$, taking the gradient of (7.4-2) gives

$$\boldsymbol{\nabla}\psi = \left(g_r\boldsymbol{e}_r - \frac{g}{2}\boldsymbol{e}_1\right)e^{-\frac{x_1}{2}},$$

and substituting this into (7.4-4) gives

$$(7.4\text{-}5) \qquad \boldsymbol{u} = e^{\frac{x_1}{2}}\left(-g_r\boldsymbol{e}_\vartheta + \frac{g}{2}\boldsymbol{e}_2\right).$$

The average of u_1 over $r = \varepsilon$ vanishes:

$$\langle u_1 \rangle = g_r(\varepsilon)\frac{1}{2\pi}\int_0^{2\pi} e^{\frac{\varepsilon}{2}\cos\vartheta}\sin\vartheta\, d\vartheta = 0.$$

The average of u_2 requires more work:

$$\langle u_2 \rangle = -g_r(\varepsilon)\frac{1}{2\pi}\int_0^{2\pi} e^{\frac{\varepsilon}{2}\cos\vartheta}\cos\vartheta\, d\vartheta + \frac{1}{2}g(\varepsilon)\frac{1}{2\pi}\int_0^{2\pi} e^{\frac{\varepsilon}{2}\cos\vartheta}\, d\vartheta,$$

or

$$(7.4\text{-}6) \qquad \langle u_2 \rangle = -g_r(\varepsilon)\left\{\frac{\varepsilon}{4} + O(\varepsilon^3)\right\} + \frac{1}{2}g(\varepsilon)\{1 + O(\varepsilon^2)\}.$$

Finally, we must extract $r \to 0$ asymptotic formulas for $g(r)$ and $g_r(r)$ from the integral representation (7.4-3). We start with the change of variable $x = re^\zeta$, like in Problem 4.9b, and (7.4-3) becomes

$$(7.4\text{-}7) \qquad g = \frac{1}{2\pi}\int_r^{\infty} e^{-\frac{1}{2}(x+\frac{r^2}{x})}\frac{dx}{x}.$$

An integration by parts advances us to

$$(7.4\text{-}8) \qquad g = -\frac{1}{2\pi}e^{-r}\log r + \frac{1}{4\pi}\int_0^{\infty} e^{-\frac{1}{2}(x+\frac{r^2}{x})}\left(1 - \frac{r^2}{x^2}\right)\log x\, dx.$$

How do we know to do *this* integration by parts? We note that the $r \to 0$ behavior of g should be like that of the point source solution $-\frac{1}{2\pi}\log r$ of

Laplace's equation in $\mathbb{R}^2$. So doing the integration by parts of (7.4-7) with $dv = \frac{dx}{x}$ gives us our logarithm. In the limit $r \to 0$ (7.4-8) reduces to

$$(7.4\text{-}9) \qquad g = -\frac{1}{2\pi}\log r + \frac{1}{4\pi}\int_0^\infty e^{-\frac{x}{2}}\log x\,dx + o(r).$$

The notation $o(r)$ denotes a truncation error that vanishes as $r \to 0$. In (7.4-9), $e^{-\frac{x}{2}}\log x$ is the formal limit of the integrand in (7.4-8) as $r \to 0$ with x fixed. It is nonuniformly valid for x comparable to r, but its contribution to the integral vanishes as $r \to 0$. In summary,

$$(7.4\text{-}10) \qquad g(r) = -\frac{1}{2\pi}\log r + \frac{\gamma}{4\pi} + o(r),$$

where

$$(7.4\text{-}11) \qquad \gamma := \int_0^\infty e^{-\frac{x}{2}}\log x\,dx$$

is the Euler constant.

We also need an appropriate approximation to $g_r(r)$ as $r \to 0$. Look back at the formula (7.4-6) for $\langle u_2 \rangle$: By (7.4-10), its second term is

$$(7.4\text{-}12) \qquad \frac{1}{2}g(\varepsilon)(1 + O(\varepsilon^2)) = -\frac{1}{4\pi}\log\varepsilon + \frac{\gamma}{8\pi} + o(\varepsilon).$$

The smallest explicit term on the right-hand side is the constant $\frac{\gamma}{8\pi}$. So we need to resolve the first term of (7.4-6), $-g_r(\varepsilon)\{\frac{\varepsilon}{4}+O(\varepsilon^3)\}$, up to a constant term, and hence $g_r(\varepsilon)$ to a $\frac{1}{\varepsilon}$ term. Differentiating the $\log r$ term in (7.4-10) with respect to r and setting $r = \varepsilon$ gives $-\frac{1}{2\pi\varepsilon}$. But could there be some strong singularity in the r-derivative of the unresolved $o(r)$ term in (7.4-10)? The simplest check is to differentiate (7.4-7) with respect to r, obtaining

$$g_r = -\frac{e^{-r}}{2\pi r} - \frac{r}{2\pi}\int_r^\infty e^{-\frac{1}{2}(x+\frac{r^2}{x})}\frac{dx}{x^2}.$$

It is easy to show that the integral is bounded in absolute value by a constant independent of r. So we have

$$g_r(\varepsilon) = -\frac{1}{2\pi\varepsilon} + O(1),$$

and now (7.4-6) reduces to

$$(7.4\text{-}13) \qquad \langle u_2 \rangle = \frac{1}{4\pi}\log\frac{1}{\varepsilon} + \frac{1}{8\pi}(\gamma + 1) + O(\varepsilon).$$

c) It seems that the vortex patch translates with the uniform velocity

$$\left\{\frac{1}{4\pi}\log\frac{1}{\varepsilon} + \frac{1}{8\pi}(\gamma + 1) + O(\varepsilon)\right\}\mathbf{e}_2.$$

Our ocean gets shallower as x_1 increases, so a *rough* physical analogy would be an impenetrable boundary wall at some $x_1 > 0$. According to the analysis of the "ground effect" in Problem 6.8, the wall induces a vortex of positive circulation to travel in the $+e_2$ direction. But notice the $\log \frac{1}{\varepsilon}$ divergence of velocity as the vortex is concentrated to a vanishingly small disk. We expect that the shallow water approximation will break down if the vortex is too concentrated. More generally, the approximation of "point" vortices in two-dimensional flow, while appealing, usually cannot be retained whenever we perturb the physics away from purely two-dimensional, incompressible ideal fluid. Even more broadly, Coulomb interactions between localized objects (such as electric charges or vortices) become infinitely strong when the objects shrink down to points or curves.

Guide to bibliography. Recommended references for this chapter are Courant & Friedrichs [7], Lighthill [18], Ockendon & Ockendon [20], Saffman [21], Stoker [22], and Whitham [24].

Lighthill [18], Stoker [22], and Whitham [24] carry out linearized analysis of free surface waves, using the velocity potential for irrotational flow and Bernoulli's equation (a first integral of Euler's equation for irrotational flow). Our choice to retain velocity components as state variables is motivated by this chapter's emphasis on shallow water theory, in which horizontal velocity components are most natural state variables. The cost is a (superficially) longer discussion of linearized theory.

Stoker [22] presents a scaling-based reduction of the full water wave equations to shallow water theory, similar to ours. There is also a tradition of informal derivations based on the hydrostatic approximation of pressure, the independence of horizontal velocity upon depth, and retention of depth dependence in the (small) vertical component. This kind of derivation is presented in Courant & Friedrichs [7] and Whitham [24].

The logarithmic divergence of vortex velocity induced by a sloping bottom as the vortex radius goes to zero is reminiscent of similar divergences in the long-standing theory of thin vortex rings, reviewed in Saffman [21].

Chapter 8

Solution of the shallow water equations

Simple waves. We begin with "one-dimensional" solutions to the shallow water equations (7.37) and (7.38) with $h = h(x := x_1, t)$ and $\boldsymbol{u} = u(x, t)\boldsymbol{e}_1$. The reduced PDEs for $h(x, t)$ and $u(x, t)$ are

$$(8.1) \qquad\qquad h_t + uh_x + u_x h = 0,$$

$$(8.2) \qquad\qquad u_t + uu_x + h_x = 0.$$

A special class of solutions, called *simple waves*, is based upon an algebraic relation between pointwise values of h and u, which we write as

$$(8.3) \qquad\qquad u = U(h).$$

Substituting (8.3) into (8.1) and (8.2) gives

$$U'h_t + (UU' + U'^2 h)h_x = 0,$$
$$U'h_t + (UU' + 1)h_x = 0.$$

Consistency of these two equations implies

$$(U')^2 h = 1,$$

from which it follows that

$$(8.4) \qquad\qquad U = U_0 + 2\sqrt{h} \quad \text{or} \quad U = U_0 - 2\sqrt{h},$$

where U_0 is a constant. Let us consider the first case in (8.4) with the plus sign. Suppose that at time $t = 0$, we have an interval of x where $h \equiv 1$ and

$u \equiv 0$; that is, undisturbed ocean. Then $U_0 = -2$, so $U(h) = 2(\sqrt{h} - 1)$ and the "volume conservation" PDE (8.1) reduces to a single PDE for h,

$$(8.5) \qquad h_t + (2(\sqrt{h} - 1)h)_x = 0, \quad \text{or} \quad h_t + (3\sqrt{h} - 2)h_x = 0.$$

The other choice of u in (8.4) leads to (8.5) with $U(h)$ replaced by $-U(h)$, which can also be obtained via the change of variable $x \to -x$ because the equations (8.1) and (8.2) are invariant under the "reflection" $x \to -x$, $u \to -u$.

We may understand the solutions of (8.5) by examining their level curves in (x, t) spacetime. Let $x = x(t)$ be a level curve along which $h \equiv h_0$, a given constant. Then

$$(8.6) \qquad 0 = \frac{d}{dt}h(x(t), t) = (h_t + \dot{x}h_x)(x(t), t).$$

From the PDE (8.5), $h_t = (2 - 3\sqrt{h})h_x$, and substitution into (8.6) gives

$$(2 - 3\sqrt{h_0} + \dot{x})h_x(x(t), t) = 0.$$

This is satisfied if

$$(8.7) \qquad \dot{x} = 3\sqrt{h_0} - 2.$$

Hence, the level curve $h(x, t) = h_0$ is a straight line in (x, t) spacetime corresponding to uniform motion at velocity $3\sqrt{h_0} - 2$. This result allows construction of the solution to the initial value problem, with $h(x, 0)$ given. Figure 8.1 illustrates this construction.

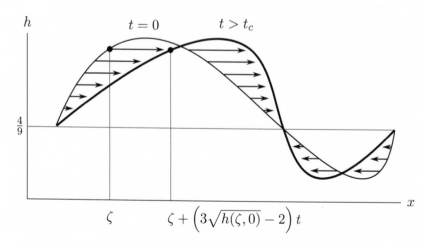

Figure 8.1.

We read off the implicit determination of $h(x, t)$ from $h(x, 0)$:

$$h(\zeta + (3\sqrt{h(\zeta, 0)} - 2)t, t) = h(\zeta, 0),$$

so

(8.8)
$$h(x,t) = h(\zeta(x,t), 0),$$

where $\zeta = \zeta(x,t)$ is determined by

(8.9)
$$x = \zeta + (3\sqrt{h(\zeta,0)} - 2)t.$$

For $h(\zeta, 0)$ positive and increasing for all ζ, the right-hand side of (8.9) is increasing in ζ, so for each (x,t), (8.9) has a unique solution for $\zeta = \zeta(x,t)$. But general initial conditions are more problematic. For instance, consider a "wavetrain" with successive "crests" and "troughs" as in Figure 8.1. Since the velocity $\dot{x} = 3\sqrt{h_0} - 2$ of the level curve $h(x,t) = h_0$ increases with h_0, each crest is closing the distance to the next trough on the right. There is a time $t = t_c$ when the graph of $h(x,t)$ versus x has a vertical tangent. This can be shown by a simple calculation: Differentiate (8.5) with respect to x to get

(8.10)
$$(h_x)_t + (3\sqrt{h} - 2)(h_x)_x = -\frac{3}{2\sqrt{h}} h_x^2.$$

Let $m(t) := -h_x(x(t), t)$, where $x(t)$ is the level curve $h(x,t) = h_0$. Then from (8.10) and $\dot{x} = 3\sqrt{h_0} - 2$ we have

$$\frac{dm}{dt} = \frac{3}{2}\frac{1}{\sqrt{h_0}} m^2.$$

The solution with given $m(0)$ is

$$m(t) = \frac{m(0)}{1 - \frac{3m(0)}{2\sqrt{h_0}}t}.$$

If $m(0) > 0$, we get $m(t) \to \infty$ as $t \to t_*^-$, where

$$t_* := \frac{2\sqrt{h_0}}{3m(0)}.$$

Hence, t_c is the minimum of the function

$$-\frac{2\sqrt{h(x,0)}}{3\,h_x(x,0)}.$$

Formal continuation of the construction in Figure 8.1 beyond $t = t_c$ leads to a multivalued "solution" for $h(x,t)$ (Figure 8.2). Of course, the shallow water equations lose validity near the event where the vertical tangent occurs, and we come to the following question: What happens to a real free surface wave as $t = t_c$ is approached, and afterwards? There *is* a "physical" continuation into $t > t_c$ within shallow water theory. To discuss it, we need a broader perspective of solutions to the PDEs (8.1)–(8.2) beyond simple waves. There is another physical reason to look beyond simple waves: The initial value problem consisting of the reduced PDE (8.5) for

h, subject to given $h(x, 0)$, looks reasonable as long as we forget about u. Given $h(x, t)$, the simple wave solution to the full equations (8.1)–(8.2) has $u(x, t) = U(h(x, t)) = 3\sqrt{h(x, t)} - 2$. In particular, the initial values of u are specifically tied to the initial values of h by $u(x, 0) = U(h(x, 0))$. From the physical perspective, "simple wave" initial data with $u = U(h)$ are highly special and artificial. It is not even clear how simple waves would arise in "generically realizable" solutions of the shallow water equations.

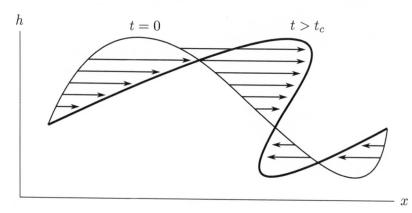

Figure 8.2.

Problem 8.1 (Simple waves in ideal fluid).   We have ideal fluid with the reduced equation of state $p = p(\rho)$, with $p'(\rho) > 0$ for stability. For one-dimensional flow with $\rho = \rho(x := x_1, t)$ and $\boldsymbol{u} = u(x, t)\boldsymbol{e}_1$, the basic equations (6.18) and (6.19) reduce to

(8.1-1) $$\rho_t + u\rho_x + u_x\rho = 0,$$

(8.1-2) $$u_t + uu_x + \frac{1}{\rho}p'(\rho)\rho_x = 0.$$

a) Determine the possible relations $u = U(\rho)$ for simple waves, assuming that $U(\rho_0) = 0$ for $\rho = \rho_0$. Given $U(\rho)$, what is the corresponding "simple wave" PDE for $\rho(x, t)$? What is the velocity $c = c(\rho)$ of a level curve of constant density ρ in (x, t) spacetime? What is $c(\rho_0)$? Explain physically.

b) Consider a right-traveling simple wave with $c(\rho) > 0$, subject to a given initial condition $\rho(x, 0)$. Derive a sufficient condition for the graph of $\rho(x, t)$ versus x to develop a vertical tangent at some time $t = t_c > 0$. For such cases, what is t_c? In *physical* gases, such a vertical tangent develops by large values of ρ propagating faster than smaller values, so that $c(\rho)$ is an *increasing* function of ρ. Derive a sufficient condition on $p(\rho)$ so that $c(\rho)$ is increasing. Show that for ideal gas, $c'(\rho) = \frac{4}{3}\frac{c(\rho)}{\rho}$.

c) We take

$$(8.1\text{-}3) \qquad\qquad \rho(x, 0) = \rho_0 + \delta\rho \cos kx,$$

where the "density amplitude" $\delta\rho$ and wavenumber k are given constants. Assuming ideal gas, evaluate the breaking time t_c explicitly in the limit $\frac{\delta\rho}{\rho_0} \to 0$. What is the distance x_c that a crest of this simple wave travels in time t_c?

Now let us be specific. A Parris Island drill instructor has sequestered his recruits deep in a long drainage culvert so he can yell at them from the open end, and the propagation of the pressure wave can be approximated by the one-dimensional model. Parris Island is at sea level, so $\rho_0 = 1.2$ kg/m^3, $c(\rho_0) = 600$ m/s, and the density amplitude $\delta\rho$ corresponds to a pressure fluctuation amplitude of 10^3 nm^{-2}. (This is only 1% of atmospheric pressure, but it is equal to 154 dB. For comparison, jet engines and lion roars top out at 120 dB.) The main temporal frequency of the yell is 100 Hz. Evaluate x_c for this "experiment".

Solution.

a) Setting $u = U(\rho)$ in (8.1-1) and (8.1-2), we obtain

$$U'\rho_t + (UU' + \rho U'^2)\rho_x = 0,$$

$$U'\rho_t + (UU' + \frac{1}{\rho}p'(\rho))\rho_x = 0,$$

and the consistency condition is

$$U'(\rho) = \pm\frac{1}{\rho}\sqrt{p'(\rho)}.$$

The integral with $U(\rho_0) = 0$ is

$$(8.1\text{-}4) \qquad\qquad U(\rho) = \pm\int_{\rho_0}^{\rho} \frac{\sqrt{p'(\sigma)}}{\sigma}\,d\sigma.$$

Given $U(\rho)$, the "simple wave" PDE for $\rho(x, t)$ is

$$(8.1\text{-}5) \qquad\qquad \rho_t + c(\rho)\rho_x = 0,$$

where $c(\rho)$ is the "level curve propagation velocity",

$$(8.1\text{-}6) \qquad\qquad c(\rho) := U(\rho) + \rho U'(\rho).$$

From (8.1-4) and (8.1-6) it follows that

$$(8.1\text{-}7) \qquad\qquad c(\rho_0) = \pm\sqrt{p'(\rho_0)}.$$

$\sqrt{p'(\rho_0)}$ is the speed of sound in fluid of uniform density ρ_0. From (8.1-5) it follows that the linearized equation for $\eta := \rho - \rho_0$ is

$$\eta_t + c(\rho_0)\eta_x = 0,$$

whose solutions are right-traveling acoustic waves for $c(\rho_0) = +\sqrt{p'(\rho_0)}$ and left-traveling acoustic waves for $c(\rho_0) = -\sqrt{p'(\rho_0)}$.

b) From the simple wave PDE (8.1-5) we derive an ODE for $m := -\rho_x$ as a function of time along a given level curve of ρ. We have

$$(\rho_x)_t + c(\rho)(\rho_x)_x = -c'(\rho)\rho_x^2,$$

so

$$\frac{dm}{dt} = c'(\rho)m^2,$$

and the solution with given $m(0)$ is

$$m(t) = \frac{m(0)}{1 - m(0)c'(\rho)t}.$$

If $m(0)c'(\rho) > 0$, $m(t)$ diverges to $+\infty$ at time

$$t = \frac{1}{m(0)c'(\rho)}.$$

For the level curve of ρ that started from $(x,0)$, $m(0) = -\rho_x(x,0)$ and $m(0)c'(\rho) = -\partial_x c(\rho(x,0))$. Hence, breaking happens if $c(\rho(x,0))$ is decreasing in some interval of x, and granted this,

$$t_c = \frac{-1}{\min \partial_x c(\rho(x,0))}.$$

From (8.1-4) with the "+" sign and (8.1-6), it follows that

$$(8.1\text{-}8) \qquad\qquad c'(\rho) = 2U' + \rho U'' = \frac{1}{2\sqrt{p'}\rho^2}(\rho^2 p')'.$$

The point of the algebraic manipulation is that $c'(\rho) > 0$ if $\rho^2 p'(\rho)$ is decreasing. For ideal gas, recall that the reduced equation of state is $p(\rho) = C\rho^{\frac{5}{3}}$, where $C > 0$ is a constant, and evaluating (8.1-8) with this $p(\rho)$ gives

$$(8.1\text{-}9) \qquad\qquad c'(\rho) = \frac{4}{3}\frac{c(\rho)}{\rho}.$$

c) We have $\min \rho_x(x,0) = -k\delta\rho$, and in the limit $\delta\rho/\rho_0 \to 0$, $c'(\rho) \to c'(\rho_0) = \frac{4}{3}\frac{c(\rho_0)}{\rho_0}$. Hence,

$$t_c = \frac{3}{4}\frac{\rho_0}{\delta\rho}\frac{1}{c(\rho)k}$$

and

$$x_c = c(\rho_0)t_c = \frac{3}{4}\frac{\rho_0}{\delta\rho}\frac{1}{k}.$$

We convert the density amplitude $\delta\rho$ into a pressure amplitude δp by $\delta p = c^2\delta\rho$, so the formula for x_c is now

$$x_c = \frac{3}{4}\frac{\rho_0 c^2(\rho_0)}{\delta p}\frac{1}{k}.$$

Inserting $\rho_0 = 1.2 \text{ kg m}^{-3}$, $c(\rho_0) = 600 \text{ m s}^{-1}$, $\delta p = 10^3 \text{ nm}^{-2}$, and $k = 1.04 \text{ m}^{-1}$ (which follows from the temporal frequency of 100 Hz), we get $x_c \simeq 312$ m.

Problem 8.2 (Longitudinal elastic waves). The nonlinear wave phenomena contained in one-dimensional ideal fluid mechanics or shallow water theory are widespread throughout nature. Here, we investigate a representative from elastic mechanics. The elastic medium at rest has uniform density ρ. Displacements of material particles from initial rest positions are in the e_1 direction and independent of x_2 and x_3. Specifically, the e_1 position of a material particle that started from rest position $x_1 = \zeta$ is represented by $x_1 = x := \zeta + \eta(\zeta,t)$. Here, $\eta(\zeta,t)$ is the displacement of the particle at time t from rest.

a) Consider the layer of medium corresponding to rest positions between ζ and $\zeta + d\zeta$. Let dx denote its thickness at time t. Compute the relative elongation $s(\zeta,t) := \frac{dx - d\zeta}{d\zeta}$, also called the *strain*, in terms of $\eta(\zeta,t)$. Explain why physical values of $s(\zeta,t)$ have $s > -1$.

The material in $x < \zeta + \eta(\zeta,t)$ experiences a *stress* or x_1-force per unit area over the plane $x_1 = \zeta + \eta(\zeta,t)$ due to the material in $x > \zeta + \eta(\zeta,t)$. This x_1-force per unit area is a function $\sigma = \sigma(s)$ of the local strain $s = s(\zeta,t)$. The basic properties reasonable of $\sigma(s)$ are $\sigma(0) = 0$ (no stress at rest) and $\sigma'(s) > 0$ for all s (stress is an increasing function of elongation). By Newton's third law, the stress acting on material in $x > \zeta + \eta(\zeta,t)$ due to material in $x < \zeta + \eta(\zeta,t)$ is $-\sigma(s(\zeta,t))$. Formulate an integral identity which expresses the time rate of change of x_1-momentum per unit area of material with $\zeta_1 < \zeta < \zeta_2$ due to stresses acting across $\zeta = \zeta_1$ and $\zeta = \zeta_2$. From this integral identity, derive the nonlinear wave equation which governs $\eta(\zeta,t)$. Formulate an equivalent pair of PDEs for the strain $s(\zeta,t)$ and particle velocity $u(\zeta,t) = \eta_t(\zeta,t)$.

b) Let $(s(\zeta,t), u(\zeta,t))$ be a solution of this pair of PDEs with a "wavefront" along some world line $\zeta = \zeta(t)$ in the (ζ,t) plane: s and u are continuous across the wavefront, but s_ζ and u_ζ can have *jump* discontinuities. What are the possible values of $\dot{\zeta}(t)$ in terms of s and u at $(\zeta = \zeta(t), t)$? The actual spatial location of the wavefront is given by $x(t) = \zeta(t) + \eta(\zeta(t), t)$. What are the possible velocities $\dot{x}(t)$?

c)   Simple elastic waves are characterized by $u = U(s)$. What are the possible $U(s)$, given $U(0) = 0$? Write down the PDE for $s(\zeta, t)$ that governs right-traveling simple waves. We analyze a *signaling problem*: Give a formal implicit expression for the right-traveling wave with $s(\zeta = 0, t) = -S(t)$, a given function.

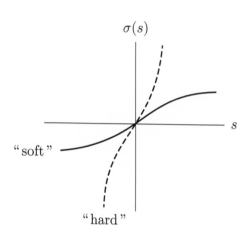

$\sigma(s)$

"soft"

s

"hard"

Figure 8.3.

d)   Assume that $S(t) \equiv 0$ for $t < 0$, and that in $t > 0$, $S(t)$ increases from $S(0) = 0$ to some steady value S_∞ as $t \to +\infty$. (Then $s(0, t)$ decreases from 0 to the *negative* value $-S_\infty$, which corresponds to a *compression* applied on $\zeta = 0$.) Show that the formal construction of a right-traveling wave in Part c gives a unique $s(\zeta, t)$ in $\zeta > 0$ for a "soft" material, which has $\sigma'(s)$ increasing in $s < 0$ (solid curve in Figure 8.3). In the case of a "hard" material, with $\sigma'(s)$ decreasing in $s < 0$ (dashed curve in Figure 8.3), show that $s(\zeta, t)$ develops a vertical tangent ($s_\zeta \to \infty$) in finite time t_c.

Solution.

a)   From $x = \zeta + \eta(\zeta, t)$, we have $dx = (1 + \eta_\zeta(\zeta, t))d\zeta$ and thus

(8.2-1) $$s(\zeta, t) := \frac{dx - d\zeta}{d\zeta} = \eta_\zeta(\zeta, t).$$

Since we cannot reverse the *ordering* of material planes perpendicular to $\boldsymbol{e}_1$, we must have $dx = (1 + s)d\zeta > 0$ for $d\zeta > 0$, and so $s > -1$.

The integral form of momentum conservation is

$$\frac{d}{dt} \int_{\zeta_1}^{\zeta_2} \rho \eta_t(\zeta, t)\, d\zeta = \sigma(s(\zeta_2, t)) - \sigma(s(\zeta_1, t)),$$

or

$$\int_{\zeta_1}^{\zeta_2} \{\rho \eta_{tt} - \partial_\zeta \sigma(\eta_\zeta)\}\, d\zeta = 0.$$

Hence, the PDE for $\eta(\zeta, t)$ is

$$\rho \eta_{tt} = \partial_\zeta \sigma(\eta_\zeta),$$

or

(8.2-2) $$\rho \eta_{tt} = \sigma'(\eta_\zeta) \eta_{\zeta\zeta}.$$

The pair of PDEs for $s = \eta_\zeta$ and $u = \eta_t$ is therefore

(8.2-3) $$s_t = u_\zeta, \quad \rho u_t = \sigma'(s) s_\zeta.$$

b) From the continuity condition

$$s(\zeta(t)^+, t) = s(\zeta(t)^-, t),$$

it follows that

(8.2-4) $$[s_t] + \dot\zeta [s_\zeta] = 0.$$

Similarly,

(8.2-5) $$[u_t] + \dot\zeta [u_\zeta] = 0.$$

From the first equation of (8.2-3), we have $[s_t] = [u_\zeta]$, and then (8.2-4) becomes

(8.2-6) $$\dot\zeta [s_\zeta] + [u_\zeta] = 0.$$

Similarly, from the second equation of (8.2-3) and (8.2-5), we find that

(8.2-7) $$\sigma'(s)[s_\zeta] + \rho \dot\zeta [u_\zeta] = 0.$$

The consistency condition associated with the homogeneous equations (8.2-6) and (8.2-7) for $[s_\zeta]$ and $[u_\zeta]$ is

$$\rho \dot\zeta^2 = \sigma'(s),$$

giving

(8.2-8) $$\dot\zeta = \pm \sqrt{\frac{\sigma'(s)}{\rho}}.$$

From $x(t) = \zeta(t) + \eta(\zeta(t), t)$, we find the actual spatial velocity of the wavefront to be

$$\dot x = \dot\zeta + \eta_t + \eta_\zeta \dot\zeta = u + (1 + s)\dot\zeta,$$

so finally,

(8.2-9) $$\dot x = u \pm (1 + s)\sqrt{\frac{\sigma'(s)}{\rho}}.$$

c) Setting $u = U(\zeta)$ in (8.2-3), we obtain

(8.2-10) $$s_t - U'(s) s_\zeta = 0,$$
$$\rho U'(s) s_t - \sigma'(s) s_\zeta = 0,$$

and the consistency condition is

$$\rho U'^2(s) = \sigma'(s),$$

or

$$U'(s) = \pm \sqrt{\frac{\sigma'(s)}{\rho}}.$$

The integral with $U(0) = 0$ is

(8.2-11) $$U(s) = \pm \int_0^s \sqrt{\frac{\sigma'(\tau)}{\rho}}\, d\tau.$$

The simple wave PDE for $s(\zeta, t)$ is the first equation of (8.2-10). For *right-traveling* simple waves, we take the *negative* $U(s)$ in (8.2-11), and so

(8.2-12) $$s_t + \sqrt{\frac{\sigma'(s)}{\rho}} s_\zeta = 0.$$

Level curves of $s(\zeta, t)$ have "velocity"

$$\dot{\zeta} = c(s) := \sqrt{\frac{\sigma'(s)}{\rho}}$$

in (ζ, t) "spacetime". Suppose we want to determine $s(\zeta, t)$ at a given event (ζ, t) in $\zeta > 0$. The level curve of s that passes through (ζ, t) originates from some $(0, \tau)$ on the t-axis, as depicted in Figure 8.4, and we have

(8.2-13) $$s(\zeta, t) = s(0, \tau) = -S(\tau).$$

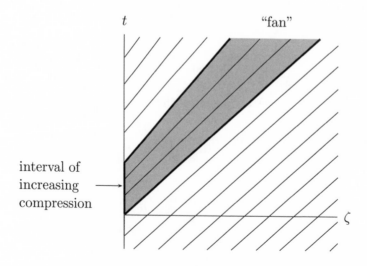

Figure 8.4.

Since the ζ-velocity of the level curve is

(8.2-14) $$\dot{\zeta} = c\,(s(0,\tau)) = v(\tau) := c\,(-S(\tau)),$$

we have

(8.2-15) $$\zeta = v(\tau)\,(t-\tau).$$

This is an implicit equation to determine $\tau = \tau(\zeta,t)$ as a function of ζ and t. Given $\tau(\zeta,t)$, the right-traveling simple wave solution is $s(\zeta,t) = s(0,\tau(\zeta,t))$.

d) The level curve of $s(\zeta,t)$ emanating from the point $(0,\tau)$ on the t-axis has velocity $\dot{\zeta} = v(\tau)$ as in (8.2-14). We have

(8.2-16) $$v'(\tau) = \frac{-\sigma''(-S(\tau))}{2\sqrt{\rho\sigma'(-S(\tau))}}S'(\tau).$$

For a "soft" material with $\sigma''(s) > 0$ in $s < 0$, the right-hand side of (8.2-16) is negative, and the level curves of $s(\zeta,t)$ "fan out" as depicted in Figure 8.4. It is clear that each point (ζ,t) in $\zeta > 0$ is crossed by exactly one level curve, so the formal construction in Part c gives a uniquely defined right-traveling simple wave.

  In the "hard" case, with $\sigma'(s)$ decreasing in $s < 0$, it follows from the simple wave PDE (8.2-12) that the value $m(t)$ of $s_\zeta(\zeta,t)$ along a level curve of s satisfies the ODE $\dot{m} = \ell m^2$, where $\ell := -\frac{\sigma''(s)}{2\sqrt{\rho\sigma'(s)}}$. We saw the same kind of argument in Problem 8.1b. Since $\sigma'(s)$ is decreasing, the coefficient ℓ is positive. For $m(0) > 0$, the solution of $\dot{m} = \ell m^2$ blows up in time $t = \frac{1}{\ell m(0)}$. For our signaling problem, are there positive $m(0)$? From the simple wave PDE (8.2-12),

$$s_\zeta = -\sqrt{\frac{\rho}{\sigma'(s)}}s_t \to \sqrt{\frac{\rho}{\sigma'(-S(t))}}S'(t) > 0$$

as $\zeta \to 0^+$. Hence, $s(\zeta,t)$ develops a vertical tangent.

Characteristics and Riemann invariants. Simple waves are characterized by

(8.11) $$R_+ := u + 2\sqrt{h}$$

or

(8.12) $$R_- := u - 2\sqrt{h}$$

equal to a uniform constant. The quantities R_+ and R_- have a basic significance in describing general solutions of the shallow water PDEs. Let us examine level curves of R_+ or R_- in spacetime. Let $x = x(t)$ be a level curve of R_+. We have

$$0 = \frac{d}{dt}R_+(x(t), t) = u_t + \dot{x}u_x + \frac{1}{\sqrt{h}}(h_t + \dot{x}h_x),$$

where it is understood that the derivatives of h and u in the right-hand side are evaluated at $(x(t), t)$. Now substitute for u_t and h_t from (8.1) and (8.2) and rearrange. We get

$$0 = (\dot{x} - u - \sqrt{h})\left(u_x + \frac{1}{\sqrt{h}}h_x\right) = (\dot{x} - u - \sqrt{h})(R_+)_x.$$

Hence, $x = x(t)$ is a level curve of R_+ if

(8.13) $\dot{x} = u + \sqrt{h}.$

Similarly, $x = x(t)$ is a level curve of R_- if

(8.14) $\dot{x} = u - \sqrt{h}.$

The families of curves defined by (8.13) or (8.14) are called *characteristics* associated with their respective *Riemann invariants* R_+ or R_-.

The characteristics and Riemann invariants inform the structure of solutions to the initial value problem, in which $h(x, 0) > 0$ and $u(x, 0)$ are given. If these initial values and their x-derivatives are bounded, then there is a unique continuous solution for h and u in some time slice $0 < t < T$. For (x, t) in this time slice, there is a unique event $(\zeta_+, 0)$ on the x-axis that is connected to (x, t) by a "+" characteristic with velocity $\dot{x} = u + \sqrt{h}$. There is a unique "−" characteristic with velocity $\dot{x} = u - \sqrt{h}$ which connects (x, t) to $(\zeta_-, 0)$ with $\zeta_- > \zeta_+$. For given h and u, ζ_+ and ζ_- are functions of (x, t). The Riemann invariants associated with the + and − characteristics are $R_+ = u + 2\sqrt{h}$ and $R_- = u - 2\sqrt{h}$. Hence, h, u are related to their initial values h_+, u_+ at $(\zeta_+, 0)$ and h_-, u_- at $(\zeta_-, 0)$ by

(8.15) $u + 2\sqrt{h} = u_+ + 2\sqrt{h_+}, \quad u - 2\sqrt{h} = u_- - 2\sqrt{h_-},$

and these can be "solved" for h and u at (x, t). Of course this "solution" is based on prior knowledge of h, u. The main point is that h, u at (x, t) are determined by initial conditions at two events $(\zeta_+, 0)$ and $(\zeta_-, 0)$ on the x-axis which are connected to (x, t) by + and − characteristics.

Now assume that the initial values $h(x, 0), u(x, 0)$ represent a localized perturbation of otherwise undisturbed ocean, so that $h(x, 0) \equiv 1$ and $u(x, 0) \equiv 0$ in $|x| > 1$. Figure 8.5 is a qualitative picture of the + and − characteristics. The + characteristic emanating from $(1, 0)$ is the line $x = t + 1$. Assuming we are in the strip $0 < t < T$ where h and u are continuous, we have $h \equiv 1$ and $u \equiv 0$ along this characteristic, hence its uniform velocity $\dot{x} = 0 + \sqrt{1} = 1$. Similarly, the − characteristic emanating

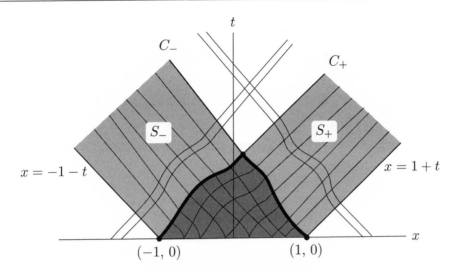

Figure 8.5.

from $(-1, 0)$ is the line $x = -t - 1$. The domain of influence of the initial data along the interval $|x| < 1$ of the x-axis is a subset of the truncated cone defined by $t > 0$, $|x| < t + 1$. Let us look inside the cone. There is a $-$ characteristic emanating from $(1, 0)$, labeled C_- in Figure 8.5. Its value of R_- is $0 - 2\sqrt{1} = -2$. The $+$ characteristics from $(x, 0)$ with $|x| < 1$ carry values of R_+ determined by initial values $h(x, 0), u(x, 0)$ in $|x| < 1$. In general, R_+ is not constant along the segment of C_- that crosses these $+$ characteristics (darkened in Figure 8.5), and nor are the values of h and u as determined by (8.15). The characteristic velocity $\dot{x} = u - \sqrt{h}$ along this "initial" segment of C_- is generally nonconstant as well. The portion of C_- beyond this initial segment has $R_+ \equiv 2$, since the "+" characteristics crossing this "later" portion of C_- come from $(x, 0)$ with $x < -1$, where $h(x, 0) \equiv 1$, $u(x, 0) \equiv 0$. The values of h, u along the "later" portion are $h \equiv 1$, $u \equiv 0$, and the velocity has constant value $\dot{x} = 0 - \sqrt{1} = -1$. Similarly, there is a $+$ characteristic C_+ emanating from $(-1, 0)$, also with an "initial" segment of nonconstant velocity and a "later" segment with $h \equiv 1, u \equiv 0$ and constant velocity $+1$. The cone of spacetime above C_+ and C_- has uniform values $h \equiv 1, u \equiv 0$ because it is covered by characteristics that come only from $(x, 0)$ with $|x| > 1$. Hence, an observer at fixed x will see an initial "storm", followed by "calm".

Two regions, labeled S_+ and S_- in Figure 8.5, remain to be accounted for. In S_+, R_- has uniform value $0 - 2\sqrt{1} = -2$, so

$$u = 2(\sqrt{h} - 1)$$

in S_+. Hence, we see that S_+ is occupied by a simple wave. Substituting $u = 2(\sqrt{h} - 1)$ into (8.1) gives the same reduced PDE (8.5) for h that we examined before. From (8.5) we found that level curves of h have velocity $3\sqrt{h} - 2$. The velocity of $+$ characteristics in S_+ is the same:

$$(8.16) \qquad \dot{x} = u + \sqrt{h} = 2(\sqrt{h} - 1) + \sqrt{h} = 3\sqrt{h} - 2.$$

Hence, $+$ characteristics in S_+ are level curves of h, and it follows from (8.16) that each of them has constant velocity. Similarly, the region S_- is occupied by a simple wave, and the $-$ characteristics in S_- are straight lines too.

Problem 8.3 (Characteristics and Riemann invariants for elastic waves). Recall the PDEs (8.2-3) for the strain $s(\zeta, t)$ and velocity field $u(\zeta, t)$ of a one-dimensional elastic wave:

$$(8.3\text{-}1) \qquad s_t = u_\zeta, \quad \rho u_t = \sigma'(s)s_\zeta.$$

a) Derive the Riemann invariants and associated characteristic velocities for these PDEs.

b) In geophysical prospecting, an elastic wave in the earth is induced by pounding the ground with an air-driven hammer. The "shot problem" is a one-space-dimension model: The PDEs (8.3-1) apply in $\zeta > 0$, and on the surface $\zeta = 0$ we have the stress as a given function of time,

$$(8.3\text{-}2) \qquad \sigma(s(0, t)) = -p(t).$$

Here, $p(t)$ is a given smooth function that is zero for all t except on an interval $(0, T)$ corresponding to the "hammer blow", in which $p(t)$ is positive. Hence, the surface stress $\sigma(s(0, t))$ is *negative*, corresponding to *compression*. The initial conditions are $s \equiv 0$, $u \equiv 0$ for $\zeta > 0$, $t < 0$. Show how the shot problem leads to an actual physical realization of the compressive simple waves examined in Problem 8.2c.

Solution.

a) Recall the construction of simple waves with $u = U(s)$. In Problem 8.2c we found that

$$U'(s) = \pm\sqrt{\frac{\sigma'(s)}{\rho}},$$

and integration gives possible functions $U(s)$,

$$(8.3\text{-}3) \qquad U(s) = R \pm \int_0^s \sqrt{\frac{\sigma'(\tau)}{\rho}}\, d\tau,$$

where R is a constant. Imitating the analysis of the shallow water equations, we surmise that our Riemann invariants are

(8.3-4) $$R_+ = u - U(s), \quad R_- = u + U(s),$$

where $U(s)$ is specifically

(8.3-5) $$U(s) := \int_0^s \sqrt{\frac{\sigma'(\tau)}{\rho}} \, d\tau,$$

obtained by taking the plus sign and $R = 0$ in (8.3-3). Let $\zeta = \zeta(t)$ be a level curve of R_+ in (ζ, t) spacetime. We calculate

$$0 = \frac{d}{dt} R_+(\zeta(t), t) = u_t + \dot\zeta u_\zeta - \sqrt{\frac{\sigma'(s)}{\rho}} (s_t + \dot\zeta s_\zeta),$$

where it is understood that u, s and their derivatives are evaluated at $(\zeta(t), t)$. Substituting for u_t and s_t from (8.3-1) and rearranging gives

$$0 = \left(\dot\zeta - \sqrt{\frac{\sigma'(s)}{\rho}} \right) (R_+)_\zeta,$$

and we discern the characteristic velocity associated with R_+,

$$\dot\zeta = \sqrt{\frac{\sigma'(s)}{\rho}} = U'(s).$$

A similar calculation derives the characteristic velocity $-U'(s)$ associated with R_-.

b) The support of u and s in the (ζ, t) plane is the sector of the first quadrant with $t > 0$ and $0 < \zeta < U'(0)t$, which is "above" the "+" characteristic $\zeta = U'(0)t$ emanating from $(\zeta, t) = (0, 0)$. Assume that u and s are smooth in the portion S of this sector with $t < T$ sufficiently small, depicted in Figure 8.6. "−" characteristics with velocity $-U'(s)$ cross into S "carrying"

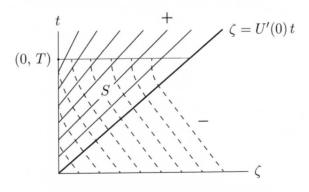

Figure 8.6.

$R_- = 0$ due to the initial conditions $u \equiv 0$, $s \equiv 0$ from $\zeta > U'(0)t$. Hence, we have $u = -U(s)$ in S, and $s(\zeta, t)$ is a simple wave satisfying the PDE $s_t + U'(s)s_\zeta = 0$, the same as (8.2-12). Assuming as we do that $\sigma = \sigma(s)$ is monotone increasing, the boundary condition (8.3-2) implies that $s(0, t)$ is some *negative* function of t in $(0, T)$, determined from $p(t)$. The boundary values $s(0, t)$ are "propagated" into S along "+" characteristics, which are straight lines with slopes $U'(s(0, t))$. Hence, $s(\zeta, t) < 0$ in S, corresponding to a *compression* wave.

Problem 8.4 (Beginning of a tsunami).   For times $t < 0$, the ocean bottom has the topography depicted in Figure 8.7. The undisturbed depth

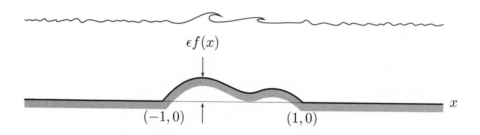

Figure 8.7.

is $1 - \varepsilon f(x)$, where $\varepsilon > 0$ is a gauge parameter and $f(x) = 0$ in $|x| > 1$, $f(x) > 0$ in $|x| < 1$. At $t = 0$ there is an "earthquake" which abruptly drops the ridge in $|x| < 1$, and we have flat ocean bottom for $t > 0$. The water above the ridge is dropped impulsively, so we expect approximate initial conditions for the shallow water PDEs:

(8.4-1) $h(x, t = 0^+) = 1 - \varepsilon f(x), \quad u(x, 0) = 0.$

The solutions for h and u in $t > 0$ take asymptotic forms

(8.4-2) $h = 1 + \varepsilon\eta(x, t, \varepsilon), \quad u = \varepsilon v(x, t, \varepsilon)$

as $\varepsilon \to 0$.

a)   Determine the leading-order approximations $\eta^0 := \eta(x, t, \varepsilon = 0)$ and $v^0 := v(x, t, 0)$ using the Riemann invariants and characteristics associated with the shallow water PDEs.

b)   Take $f(x)$ to have the qualitative form as in Figure 8.7. Construct a visualization of h at some time t after the right- and left-traveling waves have separated. Indicate the velocity field $\boldsymbol{u}$ by horizontal arrows. What is the

x-displacement of a fishing boat on the deep ocean due to the right-traveling wave passing under it?

c) To get a concrete physical sense of the "early tsunami", let us put in some plausible numbers: The ocean depth is $2000\,\mathrm{m}$, and the vertical drop of the sea floor is on the order of $10\,\mathrm{m}$ over a $10\,\mathrm{km}$-wide strip. You are in a fishing boat when the right-traveling wave passes underneath. Give estimates for: the up and down vertical displacement of the boat as the wave passes, how long it takes, the characteristic x-velocity of the water, and the total x-displacement induced by the wave.

Solution.

a) Substituting the representation (8.4-2) of h and u into (8.11) and (8.12) gives

(8.4-3)
$$r_+ := \frac{R_+ - 2}{\varepsilon} = v + \frac{2}{\varepsilon}\{\sqrt{1 + \varepsilon\eta} - 1\} = v + \eta + O(\varepsilon),$$
$$r_- := \frac{R_- + 2}{\varepsilon} = v - \frac{2}{\varepsilon}\{\sqrt{1 + \varepsilon\eta} - 1\} = v - \eta + O(\varepsilon).$$

The velocities (8.13) and (8.14) of the $+$ and $-$ characteristics in terms of η and v are

(8.4-4)
$$+: \quad \dot{x} = \varepsilon v + \sqrt{1 + \varepsilon\eta} = 1 + O(\varepsilon),$$
$$-: \quad \dot{x} = \varepsilon v - \sqrt{1 + \varepsilon\eta} = 1 + O(\varepsilon).$$

Let ζ_+ and ζ_- represent the x-intercepts of $+$ and $-$ characteristics which pass through a given event (x, t) in $t > 0$. We have

$$r_+(x, t, \varepsilon) = r_+(\zeta_+, 0, 0),$$
$$r_-(x, t, \varepsilon) = r_-(\zeta_-, 0, 0),$$

and in the limit $\varepsilon \to 0$,

(8.4-5)
$$(v^0 + \eta^0)(x, t) = (v^0 + \eta^0)(\zeta_+, 0) = -f(\zeta_+),$$
$$(v^0 - \eta^0)(x, t) = (v^0 - \eta^0)(\zeta_-, 0) = f(\zeta_-).$$

Here, we used (8.4-3) and the initial conditions on η and v induced by (8.4-1), namely $\eta^0(x, 0) = f(x)$ and $v^0(x, 0) = 0$. From (8.4-4) we see that for $\varepsilon = 0$, the velocities of the $+$ and $-$ characteristics are $+1$ and -1, so

$$\zeta_+ = x - t, \quad \zeta_- = x + t,$$

as depicted in Figure 8.8. Hence, (8.4-5) gives

(8.4-6)
$$\eta^0(x,t) = -\frac{1}{2}f(x-t) - \frac{1}{2}f(x+t),$$
$$v^0(x,t) = -\frac{1}{2}f(x-t) + \frac{1}{2}f(x+t).$$

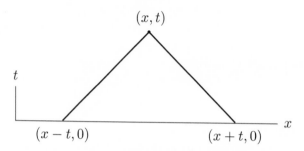

Figure 8.8.

b) Figure 8.9 is the visualization of the separated right- and left-traveling waves. We see that the velocity field in the right- (left-) traveling wave

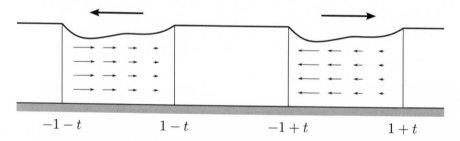

Figure 8.9.

is positive (negative). The right-traveling wave carries the fishing boat to the left as it passes underneath. The x-displacement as a function of time satisfies the ODE

(8.4-7)
$$\dot{x} = -\frac{\varepsilon}{2}f(x-t).$$

Assuming $\varepsilon \ll 1$, the total change in x is much less than 1, and we approximate x in the right-hand side as a constant. Then integration of (8.4-7) gives

$$\delta x = -\frac{\varepsilon}{2}\int_{x-1}^{x+1} f(x-t)\,dt = -\frac{\varepsilon}{2}\int_{-1}^{1} f(s)\,ds.$$

c) If we normalize the nondimensional solution so that max $f = 1$, then ε in (8.4-1) represents the *relative* uplift of the sea floor:

$$\varepsilon = 10\,\text{m} \div 2000\,\text{m} = 0.005.$$

From (8.4-6) we see that the height amplitude of the right-traveling wave is *half* of the uplift, so the rise and fall of the boat is 5 m. The wave speed in open ocean is

$$\sqrt{gH} \simeq \sqrt{(10 \,\mathrm{m/s^2})(2000 \,\mathrm{m})} \simeq 141 \,\mathrm{m/s}$$

(faster than an Olympic archery arrow traveling at approximately 70 m/s), and it takes $10^4 \,\mathrm{m} \div 141 \,\mathrm{m/s} \simeq 71 \,\mathrm{s}$ for the 10 km of wave to pass. The horizontal flow velocity is on the order of $\frac{\varepsilon}{2}\sqrt{gH} \simeq .071 \,\mathrm{m/s}$. In the 71 s it takes the wave to pass, our fishing boat is displaced horizontally by distance $\delta x \simeq -5 \,\mathrm{m}$. The whole event is hardly noticeable.

Dam break problem. This problem is a classical initial value problem of the shallow water PDEs with initial conditions $h(x,0) \equiv 1, u(x,0) \equiv 0$ in $x < 0$ and $h(x,0) \equiv 0$ in $x > 0$. Figure 8.10 shows the expected pattern of characteristics. C_+ and C_- are the $+$ and $-$ characteristics emanating from the dam break at $(x,t) = (0,0)$. The characteristic C_- propagates

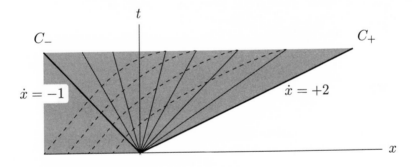

Figure 8.10.

"backwards" into formerly dammed-up ocean, with constant velocity $\dot{x} = 0 - \sqrt{1} = -1$ based on $u \equiv 0, h \equiv 1$ along C_-. The cone of spacetime between the characteristics C_- and C_+ corresponds to a flood wave, with C_+ leading the charge. This "flood wave cone" is covered by $+$ characteristics (dashed curves in Figure 8.10) emanating from points $(x,0)$ with $x < 0$. Hence, $R_+ := u + 2\sqrt{h}$ has a uniform value of 2 in the cone, so the flood wave is actually a simple wave with $u = 2(1 - \sqrt{h})$. For this simple wave, the "$-$" characteristics are level curves of h, with constant velocities

(8.17) $$\dot{x} = 2 - 3\sqrt{h}.$$

The "−" characteristics emanate from the origin $(x, t) = (0, 0)$, so the characteristic "carrying" a given value h_0 of h is

(8.18) $$x = (2 - 3\sqrt{h_0})t.$$

The values of h range from $h = 1$ along C_- to $h = 0$ along C_+. The *zero* value of h along C_+ actually makes sense: When the dam is "removed" at $(x, t) = (0, 0)$, fluid particles near the bottom of the vertical wall of water, being pressed by the layers of water above, accelerate the fastest in the $+x$ direction. Of course this initial evolution is *not* described within shallow water theory. In any case, once it is assumed that $h = 0$ along C_+, we obtain from (8.17) the velocity of the C_+ characteristic, $\dot{x} = 2$. The actual values of $h(x, t)$ in the flood wave cone are found by eliminating h_0 from equation (8.18) and

$$h(x, t) = h_0.$$

Hence,

(8.19) $$h(x, t) = \left(\frac{2}{3} - \frac{1}{3}\frac{x}{t} \right)^2.$$

The velocity u is recovered from

(8.20) $$u = 2(1 - \sqrt{h}) = \frac{2}{3}\left(1 + \frac{x}{t} \right).$$

Figure 8.11 is a visualization of the flow at time $t > 0$.

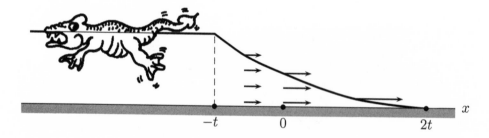

Figure 8.11.

Problem 8.5 (Wave generator).   We have water to the right of a movable vertical wall. The world line of the wall in (x, t) spacetime is $x = X(t)$, where $X(t)$ is a given function. For $t < 0$, we have $X(t) = 0$, and there is "ocean" at rest in $x > 0$, with $u = 0$ and $h = 1$. For $t > 0$, the wall executes some prescribed back and forth motion, and we expect ripples to propagate into $x > X(t)$.

a) Use Riemann invariants and characteristics to formally construct an implicit solution for h and u at events (x, t) in the "ocean" region of spacetime, $R : x > X(t)$. Your calculation will be based on the following formal assumptions: For $t > 0$, a segment of a $-$ characteristic in R connects (x, t) to the positive x-axis, and a segment of a $+$ characteristic connects (x, t) to a *unique* event on the world line of the wall. h and u are assumed to be continuous along these segments of $+$ and $-$ characteristics.

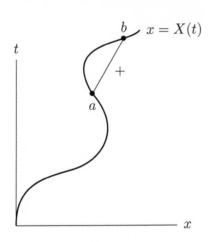

Figure 8.12.

b) We examine some basic constraints on $X(t)$ to assure that the implicit solution in Part a is well defined and well behaved for (x, t) sufficiently near the wall world line: For continuity, the restriction of h to $x = X(t)$ should approach 1 as $t \to 0^+$. Show that this forces $\dot{X}(0) = 0$. In this case, what is the region of spacetime in which h and u are affected by the motion of the wall? Show that $h > 0$ at the wall forces $\dot{X} > -2$. If $\dot{X} < -2$ for all $t > 0$, what happens to the ocean? Why do we generally object to a $+$ characteristic that "begins" at one event on the wall world line and "ends" at some later event along it, as depicted in Figure 8.12? Determine a "speed limit" c on the wall, such that $|\dot{X}| < c$ is sufficient to assure that a $+$ characteristic that "begins" on the wall world line does not end on it (i.e., no "end" event like b in Figure 8.12 can occur).

Solution.

a) Figure 8.13 depicts an event (x, t) in the $t > 0$ portion of R, together with $+$ and $-$ characteristics extrapolated from (x, t) backward in time. The $-$ characteristic is continued to the positive x-axis, where $h = 1$ and $u = 0$,

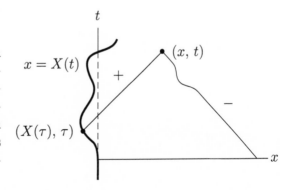

Figure 8.13.

so the value of the R_- Riemann invariant at (x, t) is

(8.5-1) $$R_- = u - 2\sqrt{h} = -2.$$

Solving (8.5-1) for h in terms of u, we get

(8.5-2) $$h = \left(1 + \frac{u}{2}\right)^2.$$

On the wall we have the kinematic boundary condition

(8.5-3) $$u(X(t), t) = \dot{X}(t),$$

and upon combining (8.5-2) and (8.5-3) we deduce that h at the wall is

(8.5-4) $$h(X(t), t) = \left(1 + \frac{\dot{X}(t)}{2}\right)^2.$$

We now look at the information "carried" by the $+$ characteristic in Figure 8.13. At some $t = \tau$, it intersects the wall world line. Given τ, we have

$$R_+(x, t) = R_+(X(\tau), \tau),$$

or

(8.5-5) $$(u + 2\sqrt{h})(x, t) = \dot{X}(\tau) + 2\left(1 + \frac{\dot{X}(\tau)}{2}\right) = 2(1 + \dot{X}(\tau)).$$

The right-hand side is $(u + 2\sqrt{h})(X(\tau), \tau)$ with $u = \dot{X}$ as in (8.5-3) and $h = (1 + \frac{\dot{X}}{2})^2$ as in (8.5-4). Solving (8.5-1) and (8.5-5) for u and h at (x, t), we find

(8.5-6)
$$u(x, t) = \dot{X}(\tau),$$
$$\sqrt{h}(x, t) = 1 + \frac{\dot{X}(\tau)}{2}.$$

It remains to identify the time τ at which the $+$ characteristic in Figure 8.13 "emerges" from the wall world line. Since R_- is a uniform constant as in (8.5-1), we have a $+$ simple wave; recall that the $+$ characteristics of a $+$ simple wave are straight lines. The constant velocity of the $+$ characteristic from $(X(\tau), \tau)$ is

(8.5-7) $$\dot{x} = u + \sqrt{h} = 1 + \frac{3}{2}\dot{X}(\tau),$$

where we have substituted for u and h from (8.5-6). Since the $+$ characteristic with this velocity connects $(X(\tau), \tau)$ and (x, t), we have

(8.5-8) $$x - X(\tau) = \left(1 + \frac{3}{2}\dot{X}(\tau)\right)(t - \tau).$$

This equation determines $\tau = \tau(x, t)$ as an implicit function of x and t, and given $\tau(x, t)$ and (8.5-6), it determines u and h at (x, t).

b) From (8.5-4), we see that $h(X(0^+), 0^+) = 1$ forces $\dot{X}(0^+) = 0$. In this case, the $+$ characteristic emanating from $(x, t) = (0, 0)$ is $x = t$, and the region of spacetime that feels the motion of the wall is $X(t) < x < t$, $t > 0$. We also see from (8.5-4) that h approaches *zero* when $\dot{X} \to -2$ from above. Physically, if $\dot{X} < -2$, the wall separates from the ocean and we have a situation, like in our flood wave example, of "ocean" spreading over newly exposed "sea floor". If we have $\dot{X} < -2$ for all $t > 0$, the support of h is $x > -2t$, and in this case we indeed have the flood wave solution.

We explain why the $+$ characteristic which begins and ends on the wall world line $x = X(t)$, like in Figure 8.12, falls outside of the construction in Part a. In this construction, we have seen that R_+ at $(X(\tau), \tau)$ is $2(1 + \dot{X}(\tau))$; so if it holds, the "beginning" and "end" values of $\dot{X}$ must be equal. However, we see that the end value generally exceeds the beginning value. We now derive the speed limit $|\dot{X}| < c$ which prevents $+$ characteristics from ending on $x = X(t)$. The velocity of the $+$ characteristic from $(X(\tau), \tau)$ is $\dot{x} = 1 + \frac{3}{2}\dot{X}(\tau)$ by (8.5-7), and given the speed restriction $|\dot{X}| < c$, we have $\dot{x} > 1 - \frac{3}{2}c$. If this lower bound on $\dot{x}$ exceeds c, then the $+$ characteristic from $(X(\tau), \tau)$ will never "end" on the wall world line at some time $t > \tau$. This is illustrated in Figure 8.14. Hence, it is sufficient to take $c < 1 - \frac{3}{2}c$, i.e., $c < \frac{2}{5}$.

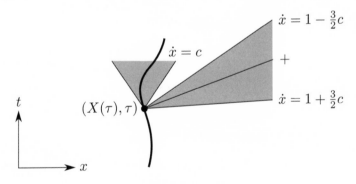

Figure 8.14.

Bores. Recall that simple wave solutions for $h(x, t)$ can "break", with the graph of h versus x developing a vertical tangent at some event (x, t). There are various scenarios for physical evolutions of the free surface after the breakdown of shallow water theory. Here is one: Look at the runoff in a flash flood channel after a shower. You see *turbulent bores* in which h

increases from one locally uniform value downstream to a larger value h' upstream as you cross a layer of churning, frothy fluid. Let us assume that the flow downstream from the bore falls within the assumptions of shallow water theory. It is not obvious that the flow upstream will be so nice, but for the purposes of a simple model, and following the analysis of Lord Rayleigh, we shall assume that shallow water theory applies in the wake of the bore as well. How are post-bore height h' and velocity u' related to pre-bore values h and u?

The shallow water PDEs can be recast in forms that express local conservation of area and momentum:

(8.21) $$h_t + (hu)_x = 0,$$

(8.22) $$(hu)_t + \left(hu^2 + \frac{h^2}{2} \right)_x = 0.$$

Note that we say "area" instead of "volume". For the one-dimensional waves we examine here, conservation of fluid volume implies conservation of area under the graph of h versus x. In (8.21) and (8.22) we recognize hu and $hu^2 + \frac{h^2}{2}$ as the fluxes of area and momentum. The local conservation laws (8.21) and (8.22) don't hold in the frothy part of the bore, but we assume that this frothy layer is not a sink of area or of momentum.

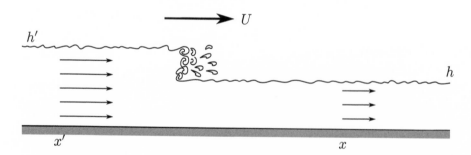

Figure 8.15.

For instance, consider the conservation of area: Let U be the bore velocity. In time dt, the area under the graph of h between x' and x in Figure 8.15 changes by amount $(h' - h)U\, dt$. Fluid enters (x', x) through x' at area per unit time $h'u'$ and leaves through x at rate hu. In time dt, the change in area is also given by $(h'u' - hu)\, dt$, and hence there is an "area" jump condition

(8.23) $$(h' - h)U = h'u' - hu.$$

Similarly, the time rate of change of momentum in (x', x) due to motion of the bore is

$$(h'u' - hu)U,$$

and this is maintained by the difference of momentum fluxes at x' and x,

$$h'u'^2 + \frac{h'^2}{2} - hu^2 - \frac{h^2}{2}.$$

Hence, there is a "momentum" jump condition

(8.24) $$(h'u' - hu)U = h'u'^2 + \frac{h'^2}{2} - hu^2 - \frac{h^2}{2}.$$

For a bore propagating into undisturbed ocean, $h \equiv 1$ and $u \equiv 0$, so the jump conditions can be reduced to

(8.25) $$u' = U\left(\frac{h'-1}{h'}\right), \quad U^2 = \frac{1}{2}h'(h'+1).$$

Hence, the bore speed $|U|$ is given in terms of h' and, given U, u' is also a function of h'. It *seems* that there is a bore for any positive h'; but there is a catch.

Let us look at energetics. There is a local conservation of energy which follows from the shallow water PDEs. The energy density is

(8.26) $$e = \frac{1}{2}hu^2 + \frac{h^2}{2}.$$

The first term is obviously kinetic energy. Problem 8.6 explains the potential energy term $h^2/2$. The flux associated with the energy density (8.26) is

(8.27) $$f = \frac{1}{2}hu^3 + h^2u.$$

As with area and momentum, local conservation of energy, expressed by $e_t + f_x = 0$, does not hold in the frothy part of the bore. However, a third, "energy" jump condition formulated in the spirit of the area and momentum jump conditions is *not* the answer—the system of three jump conditions is overdetermined. Lord Rayleigh proposed that the turbulence in the frothy bore is an energy sink. Let us examine this energy absorption in a frame of reference with the bore at the origin $x = 0$. If $x < 0$ is the "wake" of the bore, the rate of energy absorption is

(8.28) $$\omega = f' - f = \frac{1}{2}h'u'^3 + h'^2u' - \frac{1}{2}hu^3 - h^2u.$$

In (8.28), h, h', u, u' are constrained by area and momentum jump conditions with $U = 0$, that is,

(8.29) $$h'u' = hu = -\mu \text{ (common value)},$$

$$h'u'^2 + \frac{h'^2}{2} = hu^2 + \frac{h^2}{2},$$

and these may be combined into

(8.30) $$-\mu(u' - u) + \frac{1}{2}(h' + h)(h' - h) = 0.$$

With the help of (8.29) and (8.30), the right-hand side of (8.28) is reduced to a function of h, h' and μ:

$$\omega = \frac{\mu}{2}(u' + u)(u' - u) - \mu(h' - h)$$

$$= \left\{ -\frac{1}{4}(h' + h)(u' + u) - \mu \right\}(h' - h)$$

$$= \left\{ -\frac{1}{4}(h'u + hu') - \mu - \frac{1}{4}h'u' - \frac{1}{4}hu \right\}(h' - h)$$

$$= -\left\{ \frac{\mu}{2} + \frac{1}{4}(h'u + hu') \right\}(h' - h) = \frac{\mu}{4}\left\{ \frac{h'}{h} + \frac{h}{h'} - 2 \right\}(h' - h),$$

or finally,

$$(8.31) \qquad\qquad\qquad \omega = \frac{\mu}{4}\frac{(h' - h)^3}{hh'}.$$

Recall that $x < 0$ is the wake, so we need $u, u' < 0$ and hence $\mu > 0$. Given $\mu > 0$, we have $\omega > 0$ if $h' > h$. So positive energy dissipation requires the elevation to *increase* when we enter the wake.

Problem 8.6 ("Potential energy" in shallow water theory).   Consider a cylinder of water of uniform depth h and base area a. Compute the work required to slowly compress the cylinder walls against the pressure as the base area decreases from $a = a_1$ to $a = a_2 < a_1$. Identify the "potential energy" $e(h)$ per unit area so that this work equals the change in $e(h)a$.

Solution. The (dimensionless) pressure as a function of depth z is $p = z$, so the pressure force per unit length along the cylinder wall is $\int_0^h z\,dz = \frac{h^2}{2}$. The work required to compress the base area by da is $dW = \frac{h^2}{2}\,da$. Since the volume $v = ha$ of water is conserved, we have $h = v/a$, so $dW = \frac{v^2}{2}\frac{da}{a^2}$ and the work required to decrease a from a_1 to a_2 is

$$W = \frac{v^2}{2}\int_{a_2}^{a_1}\frac{da}{a^2} = \frac{v^2}{2}\left(\frac{1}{a_2} - \frac{1}{a_1} \right).$$

Let h_1 and h_2 be the depths when the base area is a_1 and a_2, respecively, so that $v = a_1 h_1 = a_2 h_2$; then we can write the work as

$$W = \frac{h_2^2}{2}a_2 - \frac{h_1^2}{2}a_1,$$

and this equals $e(h_2)a_2 - e(h_1)a_1$ for $e(h) = h^2/2$.

Problem 8.7 (Elastic shocks).   We examine weak solutions of the one-dimensional elastic wave equations (8.2-3) whose strain and velocity fields $s(\zeta, t)$ and $u(\zeta, t)$ have jump discontinuities across curves in (ζ, t) spacetime, called *shocks*. We analyze the physical basis of these shocks point by point:

a) In Problem 8.2a, $\eta(\zeta, t)$ denotes the x_1-displacement of a material particle from the rest position $x_1 = \zeta$. Recall that the strain and velocity fields are related to $\eta(\zeta, t)$ by $s = \eta_\zeta$ and $u = \eta_t$. An obvious physical requirement for a shock is that η be continuous across it. (Otherwise the shock would represent a gap or "break" in the elastic medium.) Let $\zeta = \zeta(t)$ represent the shock in (ζ, t) spacetime. Show that the jumps $[s]$ and $[u]$ of strain and velocity across the shock are related by

(8.7-1) $$[u] + \dot{\zeta}[s] = 0.$$

b)   The second physical idea is that an elastic shock is not a source of momentum. For simplicity, assume uniform pre-shock values s_+ and u_+ for strain and velocity, uniform post-shock values s_- and u_-, and uniform shock velocity $\dot{\zeta} := U$. The sign of U is such that the shock moves in the direction of the pre-shock half space. Show that

(8.7-2) $$[\sigma] + \rho\dot{\zeta}[u] = 0.$$

Here, $\sigma = \sigma(s)$ is the (given) stress-strain relation. Compute U^2 as a positive function of pre- and post-shock strains s_+ and s_-.

c)   The correct sign of U, positive or negative, is singled out by an energy dissipation principle, analogous to Lord Rayleigh's analysis of shallow water bores. From the PDEs (8.2-3) derive the local energy conservation law

(8.7-3) $$\left(\frac{1}{2}\rho u^2 + \Sigma(s)\right)_t + (-u\sigma(s))_\zeta = 0.$$

Here, $\Sigma(s)$ is the "elastic potential energy"

$$\Sigma(s) := \int_0^s \sigma(\tau)\, d\tau.$$

The total energy density, including kinetic, is $\frac{1}{2}\rho u^2 + \Sigma(s)$. In the simple case of uniform pre- and post-shock strain and velocity and uniform shock velocity, show that the shock is a *source* of energy with source strength (energy per unit time per unit area)

(8.7-4) $$\omega = -U\{\Sigma(s) - \bar{\sigma}[s]\},$$

where

$$\bar{\sigma} := \frac{1}{2}(\sigma(s^+) + \sigma(s^-)).$$

For energy dissipation, the source strength should be negative, so the signs of U and $\Sigma(s) - \bar{\sigma}[s]$ should match.

d) In Problem 8.2d, we examined the right-traveling simple wave induced by a compressive load imposed on $\zeta = 0$. In the case of "hard" material, with $\sigma''(s) < 0$ in $s < 0$, we showed that the right-traveling wave wants to form a vertical tangent. We expect the "afterlife" of a simple wave to be a weak solution for $s(\zeta, t)$ with a right-traveling shock ($U > 0$). We expect the pre- and post-shock strains s_+ and s_- to have $s_- < s_+ < 0$. Is such a shock dissipative as it should be?

Solution.

a) Differentiate the continuity condition

$$\eta(\zeta(t)^+, t) = \eta(\zeta(t)^-, t)$$

with respect to t to get $[\eta_t + \dot\zeta(t)\eta_\zeta] = 0$, and set $\eta_t = u$ and $\eta_\zeta = s$ to obtain $[u + \dot\zeta(t)s] = 0$, equivalent to (8.7-1).

Here is a formal derivation which parallels our analysis of the "area conservation" (8.23) for a shallow water bore: From $s = \eta_\zeta$ and $u = \eta_t$ we have $s_t + (-u)_\zeta = 0$, so think of s as a "density" with flux $-u$. In the simple case of uniform pre- and post-shock values of s and u and uniform shock velocity U, a "global" conservation of s in a fixed interval of ζ containing the shock is expressed by

$$-u_- - (-u_+) = U(s_- - s_+),$$

equivalent to (8.7-1).

b) The local conservation of momentum is expressed by

$$\rho u_t + (-\sigma(s))_\zeta = 0,$$

which is equivalent to the second PDE of (8.2-3). ρu is momentum density (with respect to ζ as the space variable) and $-\sigma(s)$ is its flux. The net influx of momentum into a fixed interval of ζ containing the shock,

$$-\sigma(s^-) - (-\sigma(s^+)),$$

is balanced by the rate of change

$$\rho(u_- - u_+)U$$

of momentum due to motion of the shock. Hence, $-[\sigma(s)] = \rho[u]U$, which is equivalent to (8.7-2). In (8.7-2), substitute (8.7-1) for $[u]$ (with $\dot\zeta = U$) to obtain

(8.7-5) $$U^2 = \frac{[\sigma]}{\rho[s]}.$$

Since $\sigma = \sigma(s)$ is increasing in s, $[s]$ and $[\sigma]$ have the *same* sign, and the right-hand side of (8.7-5) is positive.

c) With the help of the PDEs (8.2-3) we compute

$$\left(\frac{1}{2}\rho u^2\right)_t = \rho u u_t = u(\sigma(s))_\zeta$$
$$= (u(\sigma(s)))_\zeta - \sigma u_\zeta = (u\sigma(s))_\zeta - \sigma s_t.$$

In the right-hand side, $\sigma s_t = (\Sigma(s))_t$ and so

$$\left(\frac{1}{2}\rho u^2 + \Sigma(s)\right)_t + (-u\sigma)_\zeta = 0,$$

as in (8.7-3). Hence, the flux of energy is $-u\sigma(s)$.

The rate of change of total energy in a fixed interval of ζ containing the shock is $-U[\frac{1}{2}\rho u^2 + \Sigma(s)]$. This rate of change of energy minus the net energy influx $-u_-\sigma_- - (-u_+\sigma_+) = [u\sigma(s)]$ is the energy source strength

(8.7-6) $$\omega := -U\left[\frac{1}{2}\rho u^2 + \Sigma(s)\right] - [u\sigma(s)]$$

associated with the shock. Now it gets harder: Write (8.7-6) as

$$\omega = -U[\Sigma] - \frac{1}{2}(u_+ + u_-)\rho U(u_+ - u_-) - u_+\sigma_+ + u_-\sigma_-.$$

In the second term on the right-hand side, substitute $\rho U(u_+ - u_-) = -(\sigma_+ - \sigma_-)$ as follows from (8.7-2); this gives

$$\omega = -U[\Sigma] - \frac{1}{2}(u_+ + u_-)(\sigma_+ - \sigma_-) - u_+\sigma_+ + u_-\sigma_-$$
$$= -U[\Sigma] - \frac{1}{2}(u_+ - u_-)(\sigma_+ + \sigma_-).$$

In the second term of the right-hand side, substitute $u_+ - u_- = -U(s_+ - s_-)$ as follows from (8.7-1); so finally,

$$\omega = -U[\Sigma] + \frac{U}{2}(\sigma_+ + \sigma_-)[s],$$

which is equivalent to (8.7-4).

d) The answer is a little "graph exercise" from calculus. Look at Figure 8.16. We have $[\Sigma] = \int_{s_-}^{s_+} \sigma(s)\, ds$, so $-[\Sigma]$ is the area of the shaded region. Since $\sigma''(s) < 0$ in $s < 0$ (the "hard" case), this area is smaller than the area of the trapezoid with the same corners, so $-\bar{\sigma}[s] - (-[\Sigma]) > 0$ or $[\Sigma] - \bar{\sigma}[s] > 0$. For a right-traveling shock, $U > 0$, so we see that the energy source strength ω in (8.7-4) is negative. The shock is an energy sink.

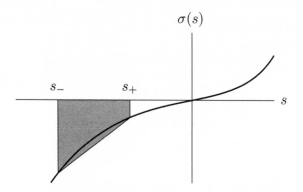

Figure 8.16.

Characteristics "eaten" by a bore. Consider a bore propagating in the $+x$ direction into undisturbed ocean with $h \equiv 1$ and $u \equiv 0$. Let $x = X(t)$ be its world line. By (8.25) its velocity is

$$(8.32) \qquad\qquad \dot{X} = \sqrt{h'\left(\frac{1+h'}{2}\right)},$$

where the prime denotes evaluation at $(x,t) = (X(t), t)$ just inside the wake. Figure 8.17 depicts a section of the bore's world line in (x,t) spacetime and the pattern of characteristics in its immediate neighborhood. Since $h' > 1$,

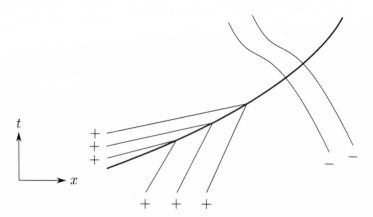

Figure 8.17.

$\dot{X} > 1$. The "$-$" characteristics cross the bore. In the undisturbed ocean where $x > X(t)$, they have constant velocity -1. Now look at the "$+$" characteristics in $x > X(t)$. They have velocity $+1$. Since $\dot{X} > 1$, these "$+$" characteristics are overtaken by the bore. Do they cross the bore and continue into the wake $x < X(t)$, like the "$-$" characteristics? No. Here is what happens: The velocity of "+" characteristics on the wake side of the bore is

(8.33) $$\dot{x} = u' + \sqrt{h'}.$$

By (8.25) and (8.32),

(8.34) $$u' = \dot{X}\frac{h' - 1}{h'} = (h' - 1)\sqrt{\frac{1 + h'}{2h'}},$$

so the "+" characteristic velocities (8.33) just inside the wake are

$$\dot{x} = \dot{X}\frac{h' - 1}{h'} + \sqrt{h'},$$

and the characteristic velocities relative to the bore velocity $\dot{X}$ are

$$\dot{x} - \dot{X} = \sqrt{h'} - \frac{\dot{X}}{h'} = \sqrt{h'} - \sqrt{\frac{1 + h'}{2h'}}.$$

The second equality follows from substituting (8.32) for $\dot{X}$. The right-hand side may be rearranged into

(8.35) $$\dot{x} - \dot{X} = (h' - 1)\frac{1 + \frac{1}{2h'}}{\sqrt{h'} + \sqrt{\frac{1+h'}{2h'}}},$$

and it is clear that $\dot{x} - \dot{X}$ is positive, so that "+" characteristics in the wake overtake the bore. In summary, the bore "absorbs" "+" characteristics from both the undisturbed ocean *and* the wake.

The afterlife of a simple wave. Recall that a localized disturbance in otherwise calm ocean can produce a pair of counter-propagating simple waves, and these can "break" by developing vertical tangents. A simple wave can be continued beyond the time of breaking by fitting in a bore. The bore world line $x = X(t)$ and solutions for h and u must be compatible with the area and momentum jump conditions. Now here is an interesting fact: A simple wave after it breaks is no longer simple. The reason is that if a Riemann invariant is uniform in the pre-bore region of spacetime, it is generally nonuniform in the wake.

We examine in detail a simple special case of a bore propagating in the $+x$ direction into undisturbed ocean, with $h \equiv 1$ and $u \equiv 0$. In the pre-bore region $x > X(t)$, R_- has uniform value $R_- = 0 - 2\sqrt{1} = -2$. The jump in R_- across the bore is

(8.36) $$R'_- - (-2) = u' + 2(1 - \sqrt{h'}) = (h' - 1)\left(\sqrt{\frac{1 + h'}{2h'}} - \frac{2}{\sqrt{h'} + 1}\right).$$

The second equality follows from substituting (8.25) for u'. It can be shown that the right-hand side of (8.36) is *positive* for $h' > 1$. In general, h' is nonuniform along the bore, so R_- does not have uniform value -2 in the wake. Hence, the wake is *not* a simple wave, even if it is the remnant of a simple wave which broke.

Propagation of a weak bore. In the limit $h' \to 1^+$, the wake is "almost" a simple wave. From Taylor expansion of (8.36) in powers of $h' - 1$ it follows that

$$(8.37) \qquad R'_- = -2 + \frac{3}{32}(h' - 1)^3 + \cdots .$$

The small change in R_- across a weak bore suggests that the dynamics of a small-amplitude "almost simple wave" with a weak bore is tractable by *perturbation analysis*. Let ε be a gauge parameter which measures $h - 1$ and u. In the limit $\varepsilon \to 0$ with x and t fixed, there are linearized right-traveling waves which propagate at velocity $+1$ with *no* change in profile. The effect of weak nonlinearity is a slow but cumulative change of wave profile over a characteristic time of duration $\frac{1}{\varepsilon}$. To capture this slow evolution of traveling waves, we introduce representations of h and u as

$$(8.38) \qquad \begin{aligned} h &= 1 + \varepsilon H(\zeta := x - t, T := \varepsilon t, \varepsilon), \\ u &= \varepsilon U(\zeta, T, \varepsilon). \end{aligned}$$

The Riemann invariant R_- in (8.12) in terms of H and U is

$$(8.39) \qquad R_- = \varepsilon U - 2\sqrt{1 + \varepsilon H} = -2 + \varepsilon(U - H) + \frac{\varepsilon^2 H^2}{4} + O(\varepsilon^3).$$

We assume a right-traveling wave propagating into undisturbed ocean, so $H, U \to 0$ as $x \to +\infty$ and the "$-$" characteristics carry uniform pre-bore value $R_- \equiv 2$; by (8.39) we have

$$(8.40) \qquad U = H - \frac{\varepsilon}{4} H^2 + O(\varepsilon^2)$$

in the pre-bore region of spacetime. In the wake, (8.37) indicates that $R_- = -2 + O(\varepsilon^3)$, so (8.40) *still* applies there.

We derive a reduced PDE for $H^0(\zeta, T) := H(\zeta, T, \varepsilon = 0)$. First, substitute the representations (8.38) into the "area" conservation equation (8.1) to obtain

$$\varepsilon^2 H_T - \varepsilon H_\zeta + \varepsilon^2 U H_\zeta + \varepsilon U_\zeta (1 + \varepsilon H) = 0,$$

or

$$(8.41) \qquad U_\zeta - H_\zeta + \varepsilon\{H_T + U H_\zeta + U_\zeta H\} = 0.$$

Now substitute for U from (8.40), and (8.41) becomes

$$H_\zeta - \frac{\varepsilon}{2}HH_\zeta - H_\zeta + \varepsilon\{H_T + 2HH_\zeta\} + O(\varepsilon^2) = 0,$$

or finally

$$H_T + \frac{3}{2}HH_\zeta = O(\varepsilon).$$

Hence, $H^0(\zeta, T)$ satisfies

(8.42) $$H_T^0 + \frac{3}{2}H^0 H_\zeta^0 = 0$$

before *and after* a possible bore.

We discuss the motion of a bore within the small-amplitude, long-time limit considered here. The "area" jump condition (8.23) in terms of the perturbation variables H and U reads

(8.43) $$(H' - H)V = (1 + \varepsilon H')U' - (1 + \varepsilon H)U.$$

Here, V is the bore velocity in (x, t) spacetime, H, U refer to values just before the bore, and H', U' refer to values just after. In (8.43), set $U = H - \frac{\varepsilon H^2}{4} + O(\varepsilon^2)$ and $U' = H' - \frac{\varepsilon H'^2}{4} + O(\varepsilon^2)$ to obtain

$$(H' - H)V = (1 + \varepsilon H')(H' - \frac{\varepsilon H'^2}{4}) - (1 + \varepsilon H)(H - \frac{\varepsilon H^2}{4}) + O(\varepsilon^2)$$

$$= H' - H + \frac{3}{4}\varepsilon(H'^2 - H^2) + O(\varepsilon^2).$$

Dividing by $H' - H$ gives

(8.44) $$V = 1 + \frac{3}{2}\varepsilon\left(\frac{H' - H}{2}\right) + O(\varepsilon^2).$$

Two remarks: First, we used the "area" jump condition explicitly, but the "momentum" jump condition didn't just disappear: It is hidden in the ε^3 smallness of R_-'s jump across the bore, which in turn validates the *asymptotic* continuity of $U - H + \frac{\varepsilon}{4}H^2$. Second, we see that the velocity of a small-amplitude bore is close to the linearized propagation velocity $+1$. The bore world line is conveniently expressed as a curve $\zeta = Z(T)$ in the (ζ, T) plane. The world line in (x, t) spacetime is

$$x = X(t) = t + Z(T),$$

and it follows that

$$V = \frac{dX}{dt} = 1 + \varepsilon\frac{dZ}{dT}.$$

Comparison with (8.44) gives

$$\frac{dZ}{dT} = \frac{3}{2}\frac{H' + H}{2} + O(\varepsilon),$$

and in the limit $\varepsilon \to 0$,

(8.45) $$\frac{dZ}{dT} \to \frac{3}{2} \frac{(H^0)' + H^0}{2}.$$

This asymptotic bore velocity can be derived by a direct (and oversimplified) appeal to "area conservation". The asymptotic PDE (8.42) is equivalent to

$$H^0_T + \left\{ \frac{3}{4}(H^0)^2 \right\}_\zeta = 0.$$

At face value, it says that H^0 (as "area per unit length") has flux $\frac{3}{4}(H^0)^2$; so in a fixed, small interval of ζ containing the bore, the "net area influx" $\frac{3}{4}(H^{0'})^2 - \frac{3}{4}(H^0)^2$ should balance the area rate of change $(H^{0'} - H^0)\frac{d\zeta'}{dT}$, and we are back to (8.45).

The reduced PDE (8.42) (called *Burger's* equation) and jump condition (8.45) comprise a famous pedagogical model of waves and shocks in nonlinear material media. Usually, this model is introduced at the very beginning as a simplified cartoon. Here, it is derived via a sequence of asymptotic reductions from the full theory of water waves. It shouldn't surprise you that "Burger's model" is an asymptotic limit of small-amplitude waves and shocks in many other contexts. For instance, you can go beyond the linearized theory of the sonic boom in Problem 6.6, or examine the "breaking" of small-amplitude but still nonlinear elastic waves in Problem 8.2.

Problem 8.8 (Area rule for Burger's shocks).   We develop a simple geometric rule for "shock fitting" in weak solutions of Burger's equation

$$u_t + uu_x = 0.$$

As we know, the implicit solution determined by propagating level curves of u in (x, t) spacetime at uniform velocities u becomes multivalued if there is an interval with $u_x(x, 0) < 0$. We propose to continue the solution past "breaking" by fitting a jump discontinuity at $x = X(t)$ as depicted in Figure 8.18.

The Z-shaped curve represents the formal multivalued solution, and the weak solution with a jump discontinuity is represented by the two thicker portions.

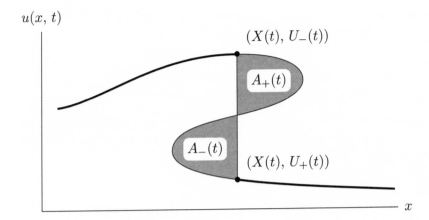

Figure 8.18.

a) Show that the propagation of the shock at velocity

(8.8-1)
$$\dot{X} = \frac{1}{2}(U_+(t) + U_-(t))$$

for all times after breaking implies equality of areas $A_+(t)$ and $A_-(t)$ in Figure 8.18.

b) Compute the solution of Burger's equation with the "triangle" initial condition

$$u(x,0) = \left[\begin{array}{ll} 0, & |x| > 1, \\ \\ 1 - |x|, & |x| < 1 \end{array} \right.$$

for $t > 0$. Compute the weak solution past breaking by fitting a shock according to the area rule. Sketch a spacetime diagram with the level curve of u and the shock.

Solution.

a) Let $C(t)$ represent the portion of the Z-shaped curve in Figure 8.18 from $(X(t), U_+(t))$ to $(X(t), U_-(t))$. The area difference $A_+ - A_-$ is

(8.8-2)
$$A(t) := \int_{C(t)} x \, du - (U_-(t) - U_+(t))X(t).$$

The line integral on the right-hand side asks for a suitable parametric representation of $C(t)$: We assume $C(t)$ is not a graph of u versus x, but in general it is not a graph of x versus u either.

The implicit solution for u based on propagating level curves of u at velocities u can be written as

$$
\begin{aligned}
x &= x(\zeta, t) := \zeta + v(\zeta)t, \\
u &= v(\zeta).
\end{aligned}
$$
(8.8-3)

Recall that ζ labels level curves of u in the (x, t) plane, and $v(\zeta)$ obviously comes from the initial condition at $t = 0$. We use the implicit solution (8.8-3) because it is a *ready-made* parametric representation of $C(t)$, with ζ in the interval $\zeta_-(t) < \zeta < \zeta_+(t)$, such that

$$
\begin{aligned}
x(\zeta_+, t) &= x(\zeta_-, t) = X(t), \\
v(\zeta_+(t)) &= U_+(t), \quad v(\zeta_-(t)) = U_-(t).
\end{aligned}
$$
(8.8-4)

Writing the line integral in (8.8-2) in parametric form, we have

$$
A(t) = \int_{\zeta_+(t)}^{\zeta_-(t)} x(\zeta, t) v'(\zeta)\, d\zeta - (U_-(t) - U_+(t)) X(t).
$$

The time derivative is

$$
\dot{A}(t) = \int_{\zeta_+}^{\zeta_-} x_t(\zeta, t) v'(\zeta)\, d\zeta
$$

$$
+ (xv')(\zeta_-, t)\dot{\zeta}_- - (xv')(\zeta_+, t)\dot{\zeta}_+ - \frac{d}{dt}\{(U_- - U_+)X\}.
$$

Using $x_t(\zeta, t) = v(\zeta)$ (from (8.8-3)) and (8.8-4), this reduces to

$$
\dot{A} = \frac{1}{2}(U_-^2 - U_+^2) - (U_- - U_+)\dot{X},
$$

and for $\dot{X} = \frac{1}{2}(U_+ - U_-)$, the right-hand side vanishes. Hence, $A_+(t) - A_-(t) = A$ is a constant in time. At the moment of breaking, A_+ and A_- are both zero, so the constant is in fact zero and therefore $A_+(t) = A_-(t)$ after breaking.

b) The initial condition at $t = 0$ is piecewise linear in x, and so is $u(x, t)$ for $t > 0$. This is easily seen by substituting linear $v(\zeta)$ in (8.8-3). Hence, "solving the initial value problem" simply means "moving the vertices of the triangle". In Figure 8.19, the vertices $(1, 0)$ and $(-1, 0)$ don't move ($x = -1, 1$ are level curves of u with value zero), and the vertex $(s(t), 1)$ has $\dot{s} = 1$ so that $s(t) = t$ since $s(0) = 0$. Hence, for $0 < t < 1$ we see a single-valued "triangle wave" whose right edge steepens to become vertical at time $t = 1$. For $t > 1$ we formally get a "multivalued triangle wave" as in the bottom graph of Figure 8.19. The shock is to be placed at $x = X(t)$ so that the areas of the shaded triangles are equal, that is,

$$
\frac{1}{2}(X - 1)\left(\frac{X - 1}{t - 1}\right) = \frac{1}{2}\left\{\frac{X + 1}{t + 1} - \frac{X - 1}{t - 1}\right\}(t - X),
$$

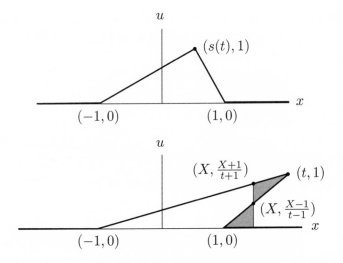

Figure 8.19.

which simplifies to

(8.8-5)
$$X(t) = 1 + \frac{t-1}{1 + \sqrt{\frac{1+t}{2}}}.$$

The pre-shock value of u is $U_- = 0$, and the post-shock value is $U_+ = \frac{X+1}{t+1}$ or, with help of (8.8-5),

$$U_+ = \left(1 + \frac{t}{1 + \sqrt{\frac{1+t}{2}}}\right) \frac{1}{t+1} \sim \sqrt{\frac{2}{t}}$$

as $t \to \infty$. Figure 8.20 is the spacetime diagram of u level curves and the shock:

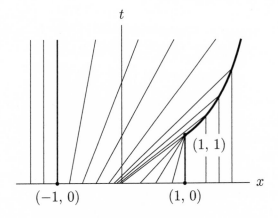

Figure 8.20.

Problem 8.9 (Traveling waves in variable-depth ocean). In Problem 7.2a we derived the modified shallow water equations for ocean of variable undisturbed depth $\ell(\boldsymbol{x})$. The one-dimensional reduced equations for surface elevation $h(x,t)$ and $\boldsymbol{e}_1$-velocity $u(x,t)$ are

$$(8.9\text{-}1) \qquad\qquad h_t + \{(\ell + h)u\}_x = 0,$$

$$(8.9\text{-}2) \qquad\qquad u_t + uu_x + h_x = 0.$$

The analysis of traveling waves proceeds one step at a time:

a) First, show that the linearized propagation velocity in uniform-depth ocean with ℓ constant is $\sqrt{\ell}$.

b) We replace the uniform bottom by "slowly varying" bottom topography, with $\ell = \ell(X := \varepsilon x)$. Here $\varepsilon > 0$ is a gauge parameter and we will analyze a certain $\varepsilon \to 0$ limit. Specifically, we examine linearized right-traveling waves localized about a world line in (x,t) spacetime whose velocity $\frac{dx}{dt}$ agrees with the local value of $\sqrt{\ell}$. It is most convenient to represent the world line by

$$(8.9\text{-}3) \qquad\qquad X = S(T := \varepsilon t).$$

Explain physically why the scaling of time with ε to make a "slow time" $T := \varepsilon t$ is compatible with the scaling of x in the "slow space" variable $X := \varepsilon x$. Can you anticipate what the ODE for $S(T)$ is going to be?

c) The actual ansatz for linearized right-traveling waves in the variable-bottom ocean is

$$(8.9\text{-}4) \qquad h = H(\zeta := x - \frac{1}{\varepsilon}S(T), T, \varepsilon), \quad u = U(\zeta, T, \varepsilon).$$

Substitute (8.9-4) for h and u into the linearizations of (8.9-1) and (8.9-2) with $\ell = \ell(X)$. By analyzing the limit $\varepsilon \to 0$, *derive* the ODE for $S(T)$ proposed in Part b, and determine general solutions for $H^0(\zeta, T) := H(\zeta, T, \varepsilon = 0)$ and $U^0(\zeta, T)$ with $H^0, U^0 \to 0$ as $\varepsilon \to +\infty$. How do the horizontal width, height, and fluid velocity of the wave respond to the bottom variation?

d) We introduce nonlinearity: The ansatz (8.9-4) is modified to

$$(8.9\text{-}5) \qquad\qquad h = \varepsilon H(\zeta, T, \varepsilon), \quad u = \varepsilon U(\zeta, T, \varepsilon).$$

Here ζ and T are the same as in Part c, and we shall see that the scaling of h and u with ε leads to an order-of-magnitude balance between nonlinearity and variable-bottom effects. Substitute (8.9-5) for h and u into the *full* PDEs (8.9-1) and (8.9-2), and modify the analysis of Part c to derive a reduced PDE for $H^0(\zeta, T)$. We can "mod out" the linear terms in this PDE by introducing the representation

$$H^0 = a(T)F\left(\eta := \frac{\zeta}{b(T)}, T\right).$$

The analysis of Part c tells us what $a(T)$ and $b(T)$ should be. Determine them. Derive the PDE for $F(\eta, T)$.

Solution.

a) The linearizations of (8.9-1) and (8.9-2) with *uniform* ℓ are

$$h_t + \ell u_x = 0, \quad u_t + h_x = 0.$$

Substituting

$$h = H(\zeta := x - ct), \quad u = U(\zeta),$$

we find

$$-cH'(\zeta) + \ell U'(\zeta) = 0, \quad H'(\zeta) - cU'(\zeta) = 0,$$

or

(8.9-6)
$$\begin{bmatrix} -c & \ell \\ 1 & -c \end{bmatrix} \begin{bmatrix} H'(\zeta) \\ U'(\zeta) \end{bmatrix} = 0,$$

and this linear homogeneous pair of equations for $H'(\zeta)$ and $U'(\zeta)$ has nonzero solutions if $c^2 = \ell$. For *right-traveling* waves, we have propagation velocity $c = +\sqrt{\ell}$. In this case, (8.9-6) implies $U'(\zeta) = \frac{1}{\sqrt{\ell}} H'(\zeta)$. Integrating with respect to ζ subject to $H, U \to 0$ as $\zeta \to +\infty$ gives $U(\zeta) = \frac{1}{\sqrt{\ell}} H(\zeta)$. Hence, the general right-traveling wave is

$$h = H(x - \sqrt{\ell} t), \quad u = \frac{1}{\sqrt{\ell}} H(x - \sqrt{\ell} t).$$

b) Since the propagation velocity is $O(1)$ in ε (its order of magnitude is independent of ε), the characteristic length and time associated with the world line should have the *same* scaling with ε. The characteristic length is presumably going to be $\frac{1}{\varepsilon}$, just like the bottom topography. Hence, the characteristic time is $\frac{1}{\varepsilon}$ too, and this is the motivation for the "slow time" $T := \varepsilon t$. The world line velocity $\frac{dx}{dt} = \frac{dX}{dT} = S'(T)$ should match the local propagation velocity $\sqrt{\ell(S(T))}$, so the ODE for $S(T)$ should be

(8.9-7)
$$S'(T) = \sqrt{\ell(S(T))}.$$

c) The linearized equations in Part a are modified to

(8.9-8)
$$h_t + \ell(X)u_x + \varepsilon \ell_X(X)u = 0,$$
$$u_t + h_x = 0.$$

Substituting into (8.9-8) the ansatz (8.9-4) for h and u, we get

(8.9-9)
$$\varepsilon H_T - S' H_\zeta + \ell(S + \varepsilon\zeta)U_\zeta + \varepsilon \ell_X(S + \varepsilon\zeta)U = 0,$$
$$\varepsilon U_T - S' U_\zeta + H_\zeta = 0.$$

The reduced $\varepsilon = 0$ equations are

$$-S'H_\zeta^0 + \ell(S)U_\zeta^0 = 0, \quad H_\zeta^0 - S'U_\zeta^0 - 0,$$

i.e., the same pair of linear equations as seen in Part a, with S' in place of c. The solvability condition is $(S')^2 = \ell(S)$, and for right-traveling waves we get the ODE $S' = +\sqrt{\ell(S)}$ as proposed in Part b. Substituting $\sqrt{\ell(S)}$ for S' in (8.9-9) and rearranging, we get

$$(8.9\text{-}10) \qquad \begin{aligned} &-\sqrt{\ell(S)}H_\zeta + \ell(S)U_\zeta + \varepsilon\{H_T + \ell_X(S)(\zeta U_\zeta + U)\} + O(\varepsilon^2) = 0, \\ &\sqrt{\ell(S)}H_\zeta - \ell(S)U_\zeta + \varepsilon\sqrt{\ell(S)}U_T = 0. \end{aligned}$$

The $\varepsilon \to 0$ limit of *either* equation gives

$$U_\zeta^0 = \frac{1}{\sqrt{\ell(S)}}H_\zeta^0,$$

and after ntegrating with respect to ζ subject to $H^0, U^0 \to 0$ as $\zeta \to +\infty$, we obtain

$$(8.9\text{-}11) \qquad\qquad U^0 = \frac{1}{\sqrt{\ell}}H^0.$$

We recall this result from Part a as well.

Adding the two equations of (8.9-10) and taking the limit $\varepsilon \to 0$ gives

$$H_T^0 + \sqrt{\ell(S)}U_T^0 + \ell_X(S)(\zeta U^0)_\zeta = 0.$$

Next, substitute for U^0 from (8.9-11), and we obtain a reduced PDE for H^0:

$$(8.9\text{-}12) \qquad 2H_T^0 - \frac{1}{2}\frac{\ell_X(S)}{\ell(S)}S'H^0 + \frac{\ell_X(S)}{\sqrt{\ell(S)}}(\zeta H^0)_\zeta = 0.$$

We express the coefficients in terms of time derivatives of $S(T)$: Differentiate the ODE (8.9-7) for $S(T)$ with respect to T to find

$$\frac{S''}{S'} = \frac{1}{2}\frac{\ell_X(S)}{\sqrt{\ell(S)}} = \frac{1}{2}\frac{\ell_X(S)}{\ell(S)}S',$$

and (8.9-12) reduces to

$$2H_T^0 + \frac{S''}{S'}(-H^0 + 2(\zeta H^0)_\zeta) = 0,$$

or finally

$$(8.9\text{-}13) \qquad\qquad H_T^0 + \frac{S''}{S'}\zeta H_\zeta^0 = -\frac{1}{2}\frac{S''}{S'}H^0.$$

The general solution of (8.9-13) is remarkably simple: just vertical and horizontal scaling of an arbitrary "profile function" $F(\cdot)$. Specifically, we seek

$$(8.9\text{-}14) \qquad\qquad H^0(\zeta, T) = a(T)F\left(\frac{\zeta}{b(T)}\right),$$

where the "vertical" and "horizontal" scale factors $a(T)$ and $b(T)$ are to be determined so that (8.9-14) satisfies (8.9-13). We find ODEs for a and b, $\frac{2a'}{a} = \frac{b'}{b} = -\frac{S''}{S'}$, which give $a = \frac{1}{\sqrt{S'}} = \ell^{-\frac{1}{4}}$ and $b = \frac{1}{S'} = \ell^{-\frac{1}{2}}$ modulo multiplicative constants. A constant in a is redundant because of the linearity of (8.9-13), and a constant in b can be absorbed by the profile function $F(\cdot)$. In summary, the general linearized right-traveling wave has the $\varepsilon \to 0$ limit

(8.9-15)
$$h \to H^0(\zeta, T) = \frac{1}{\ell(S)^{\frac{1}{4}}} F\left(\frac{\zeta}{\sqrt{\ell(S)}}\right),$$

$$u \to U^0(\zeta, T) = \frac{1}{\ell(S)^{\frac{1}{4}}} H^0(\zeta, T) = \frac{1}{\ell(S)^{\frac{3}{4}}} F\left(\frac{\zeta}{\sqrt{\ell(S)}}\right).$$

We see that the horizontal width of the wave is proportional to the local value of $\sqrt{\ell}$, the propagation velocity. The height and fluid velocity are proportional to $\ell^{-\frac{1}{4}}$ and $\ell^{-\frac{3}{4}}$.

These simple results cry out for elementary physical arguments. Are there in fact simple explanations? Yes and no. Let us do the "yes" part first: Figure 8.21 depicts the (alleged) support of some right-traveling wave in (X, T) spacetime between "leading" world line $X = S_+(T)$ and "trailing" world line $X = S_-(T)$. Presumably both $S_+(T)$ and $S_-(T)$ satisfy the "propagation velocity" ODE $S' = \sqrt{\ell(S)}$ as in (8.9-7). The wave's interval of support in X at any T has length $\delta S(T) := S_+(T) - S_-(T)$. In the limit $\delta S \to 0$, δS satisfies the *variational* ODE

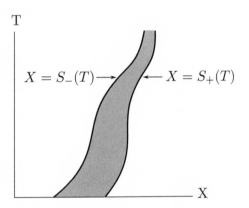

Figure 8.21.

(8.9-16)
$$(\delta S)' = \frac{1}{2}\frac{\ell_X(S)}{\sqrt{\ell(S)}}\delta S = \frac{S''}{S'}\delta S.$$

The last equality uses $\frac{1}{2}\frac{\ell_X(S)}{\sqrt{\ell(S)}} = \frac{S''}{S'}$, derived in an "accessory" calculation in Part c. It follows from (8.9-16) that

$$\delta S \propto S' = \sqrt{\ell}.$$

The usual "layman's explanation" for the rising of a wave as it nears the shore is: "All that energy is compressed into a smaller volume of water." The variable-depth shallow water PDEs (8.9-1)–(8.9-2) have a local energy conservation

$$\left\{\frac{1}{2}(\ell+h)u^2 + \frac{1}{2}h^2\right\}_t + \left\{-\left(\frac{u^2}{2}+h\right)(\ell+h)u\right\} = 0.$$

The energy density is

$$e := \frac{1}{2}(\ell+h)u^2 + \frac{1}{2}h^2.$$

In the limit $h \ll \ell$ it reduces to

$$e \sim \frac{1}{2}\ell u^2 + \frac{1}{2}h^2.$$

An order-of-magnitude balance between the kinetic and potential energies $\frac{1}{2}\ell u^2$ and $\frac{1}{2}h^2$ indicates that $\frac{h}{u}$ scales like $\sqrt{\ell}$. We saw this in the very beginning (Part a). Assuming that the interval of support of h, u, and hence e scales like $\sqrt{\ell}$, conservation of energy suggests that e scales like $\frac{1}{\sqrt{\ell}}$. We now surmise scalings of h and u: $h^2 \propto \frac{1}{\sqrt{\ell}}$, so $h \propto \ell^{-\frac{1}{4}}$ and $u \propto \ell^{-\frac{1}{2}}h \propto \ell^{-\frac{3}{4}}$, as in (8.9-15).

But before continuing, think of this: (8.9-1) says that h is locally conserved, so why don't we get $h \propto \frac{1}{\sqrt{\ell}}$ instead of $h \propto \ell^{-\frac{1}{4}}$? This is because the assumed support of h and u (Figure 8.21) is not quite right. Briefly, the right-traveling wave interacts with the varying bottom topography to generate a left-traveling wave in its "wake", $X < S_-(T)$. In the wake, h and u have amplitude ε relative to the right-traveling wave, so the wake is not seen at the order of approximation in Part c. But the width of the wake in x is on the order of $\frac{1}{\varepsilon}$, wide enough to "hide" extra $\int h\,dx$. Can the wake "hide" enough energy as well, so as to invalidate the little energy argument? No. In the wake, $e \propto h^2$ is of order ε^2 relative to e in the right-traveling wave, so the wake contribution to $\int e\,dx$ is of order ε.

d) Substituting (8.9-5) for h and u into (8.9-1)–(8.9-2) and rearranging gives

$$(8.9\text{-}17)\qquad \begin{aligned} -\sqrt{\ell}H_\zeta + \ell U_\zeta + \varepsilon\{H_T + \ell_X(S)(\zeta U)_\zeta + (HU)_\zeta\} + O(\varepsilon^2) &= 0,\\ \sqrt{\ell}H_\zeta - \ell U_\zeta + \varepsilon\sqrt{\ell}\{U_T + UU_\zeta\} &= 0. \end{aligned}$$

These are just like (8.9-10) in Part c, but with nonlinear terms. We redo the analysis of Part c. We have $U^0 = \frac{1}{\sqrt{\ell}}H^0$ as before. Adding the equations of (8.9-17) and taking the limit $\varepsilon \to 0$ gives

$$H_T^0 + \sqrt{\ell}U_T^0 + \ell_X(S)(\zeta U^0)_\zeta + \left(H^0 U^0 + \frac{\sqrt{\ell}}{2}U^{0^2}\right)_\zeta = 0.$$

Next, substitute $U^0 = \frac{1}{\sqrt{\ell}} H^0$ and use $\frac{S''}{S'} = \frac{1}{2} \frac{\ell_X(S)}{\sqrt{\ell(S)}}$ as in Part c to derive a reduced PDE for H^0,

$$(8.9\text{-}18) \qquad H^0_T + \left(\frac{S''}{S'} \zeta + \frac{3}{2S'} H^0 \right) H^0_\zeta = -\frac{1}{2} \frac{S''}{S'} H^0.$$

This is (8.9-13) with an additional nonlinear term. From the general solution (8.9-15) of the linearized equation for H^0, we propose the representation

$$(8.9\text{-}19) \quad H^0(\zeta, T) = \frac{1}{\ell(S)^{\frac{1}{4}}} F\left(\eta := \frac{\zeta}{\sqrt{\ell(S)}}, T \right) = \frac{1}{\sqrt{S'}} F\left(\eta := \frac{\zeta}{S'}, T \right).$$

Substituting (8.9-19) into (8.9-18) gives the PDE for F,

$$(8.9\text{-}20) \qquad\qquad F_T + \frac{3}{2} \frac{1}{\ell^{\frac{5}{4}}} F F_\eta = 0.$$

Problem 8.10 (Tsunami reveals its true self). We apply the theory of Problem 8.9 to describe the tsunami approaching the shore. The undisturbed ocean depth decreases linearly from the "deep ocean" value H to zero over a distance Λ much greater than the length L of the incident "deep ocean" tsunami. Accordingly, the dimensionless depth function $\ell(X)$ in Problem 8.9 is

$$(8.10\text{-}1) \qquad\qquad \ell(X) = \begin{bmatrix} 1, & X < 0, \\ 1 - X, & 0 < X < 1. \end{bmatrix}$$

Here, ℓ is in units of H and $X := \varepsilon x$, with x in units of L, so $\varepsilon = \frac{L}{\Lambda}$.

a) The leading edge of the tsunami has world line $X = S(T := \varepsilon t)$, where $S(T)$ satisfies the ODE (8.9-7). Determine $S(T)$ subject to $S(0) = 0$.

b) In the dimensionless variables of Problem 8.9, the free surface elevation h and x-velocity u associated with the tsunami wave are

$$(8.10\text{-}2) \qquad \begin{aligned} h &\sim \frac{\varepsilon}{\ell^{\frac{1}{4}}} F\left(\eta := \frac{x - \frac{S}{\varepsilon}}{\sqrt{\ell}}, T \right), \\ u &\sim \frac{\varepsilon}{\ell^{\frac{3}{4}}} F(\eta, T). \end{aligned}$$

Here, S and ℓ denote $S(T)$ and $\ell(S(T))$, and $F(\eta, T)$ satisfies the Burger's equation

$$(8.10\text{-}3) \qquad\qquad F_T + \frac{3}{2} \frac{1}{\ell^{\frac{5}{4}}} F F_\eta = 0.$$

The initial condition at $T = 0$ is based on the incident "early tsunami" which we examined in Problem 8.4. Recall that the early tsunami is a traveling *depression* of ocean surface induced by a collapse of the sea floor. For definiteness, we will take $F(\eta, 0)$ to be the "triangular trough" depicted in Figure 8.22.

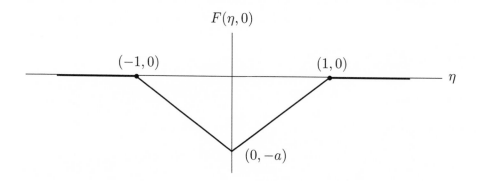

Figure 8.22.

Here, a is a scaled initial depth of the trough. Compute the solution of (8.10-3) with this initial condition up to "breaking" (when one of the triangle "walls" becomes vertical). At "breaking", what are the local undisturbed depth ℓ of ocean and the depth and width of the trough relative to the initial values at $T = 0$? Construct a visualization of the ocean surface and x-velocity field of the tsunami at breaking.

c) We want to see some actual physical magnitudes: Recall from Problem 8.4 that the deep ocean before the incline towards shore has depth $H = 2$ km, and that the "early tsunami" had a length of 10 km, with initial trough depth 5 m (hardly noticed over 10 km). The length of incline from deep ocean to shore is $\Lambda = 100$ km. Now give concrete dimensional values for the depth and width of the trough at breaking, the local depth of ocean in the wake of this (just created) bore, and the "undertow", the negative x-velocity of water just before the bore. Explain the undertow by "area conservation".

Solution.

a) The ODE for $S(T)$ is

$$\frac{dS}{dT} = \left[\begin{array}{ll} 1, & S < 0, \\ \sqrt{1-S}, & 0 < S < 1, \end{array} \right.$$

and the solution with $S(0) = 0$ is

$$S(T) = \left[\begin{array}{ll} T, & T < 0, \\ 1 - \left(1 - \frac{T}{2}\right)^2, & 0 < T < 2. \end{array} \right.$$

Figure 8.23 is the graph of the world line $X = S(T)$.

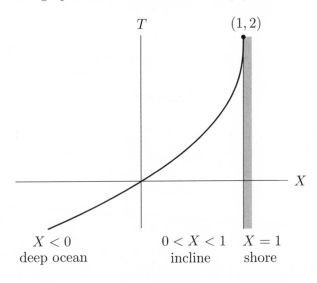

Figure 8.23.

We see the slowing of the tsunami as it approaches the shore $X = 1$ at $T = 2$. For later reference we record the depth ℓ at $X = S(T)$ as a function of T,

$$(8.10\text{-}4) \qquad\qquad \ell = 1 - S = \left(1 - \frac{T}{2}\right)^2.$$

b) Following the lead of Problem 8.8b, the solution is a "triangle wave" as in Figure 8.24 with fixed vertices at $(1,0)$ and $(-1,0)$, and a vertex $(\eta(t), -a)$ which moves to the left according to

$$\dot{\eta} = -\frac{3}{2}\frac{a}{\ell^{\frac{5}{4}}} = -\frac{3}{2}a\left(1 - \frac{T}{2}\right)^{-\frac{5}{2}}.$$

The left wall becomes vertical at $T = T_c$, so

$$(8.10\text{-}5) \qquad 1 = \frac{3}{2}a \int_0^{T_c} \left(1 - \frac{T}{2}\right)^{-\frac{5}{2}} dT = 2a\left\{\left(1 - \frac{T_c}{2}\right)^{-\frac{3}{2}} - 1\right\}.$$

The local depth $\ell = \ell_c$ at breaking follows from (8.10-4) and (8.10-5):

$$(8.10\text{-}6) \qquad\qquad \ell_c = \left(1 - \frac{T_c}{2}\right)^2 = \frac{1}{(1 + \frac{1}{2a})^{\frac{4}{3}}}.$$

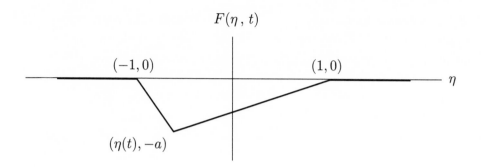

$$F(\eta, t)$$

$$(-1, 0) \qquad (1, 0)$$

$$\eta$$

$$(\eta(t), -a)$$

Figure 8.24.

The depth and width of the trough at breaking are discerned from (8.10-2). The depth relative to initial value a is

$$\ell_c^{-\frac{1}{4}} = \left(1 + \frac{1}{2a}\right)^{\frac{1}{3}}.$$

The width is compressed by factor

$$\sqrt{\ell_c} = \left(1 + \frac{1}{2a}\right)^{-\frac{2}{3}}.$$

Here is the visualization:

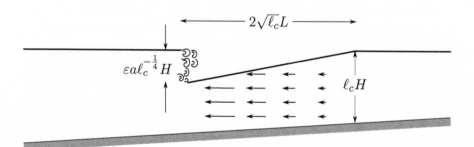

$$2\sqrt{\ell_c}L$$

$$\varepsilon a \ell_c^{-\frac{1}{4}} H$$

$$\ell_c H$$

Figure 8.25.

c) In Figure 8.25, we see that the dimensional ocean depth is $\ell_c H$, the tsunami trough depth is $\varepsilon a \ell_c^{-\frac{1}{4}} H$, and its width is $2\sqrt{\ell_c}L$. We know that $H = 2$ km, $L = 10$ km, and (from Part a) $\varepsilon := \frac{L}{\Lambda} = 10$ km $\div 100$ km $= 0.1$. It remains to determine the initial scaled trough depth, which then determines ℓ_c from (8.10-6). The initial dimensional trough depth is 5 km. In units of H, this works out to 5 m $\div$ 2000 m $= 0.0025$. Since a is this initial dimensionless depth in units of ε, we have $a = 0.0025 \div (0.1) = 0.25$. From

(8.10-6) it now follows that $\ell_c \simeq 0.01726$. Hence, the local ocean depth where the tsunami breaks is $\ell_c H \simeq 34.5$ m. The trough depth at breaking is bigger than the original 5 m by a factor of $\ell_c^{-\frac{1}{4}} \simeq 2.759$. Hence, that vertical wall of water in Figure 8.25 is 13.8 m high. The width of the trough before this (just created) bore is compressed from the original 10 km by a factor of $\sqrt{\ell_c} \simeq 0.1314$, down to 1.314 km.

As we see in Figure 8.25, the x-velocity field in this "precursor" trough before the bore is negative. We estimate this "undertow" velocity just before entering the bore in two ways:

(1) The x-velocity field in the middle of the "early tsunami" trough in deep ocean was estimated in Problem 8.4 to be -0.353 m/s. According to the asymptotic theory of Problem 8.9, this velocity amplifies by a factor of $\ell_c^{-\frac{3}{4}}$ at breaking, so we get $u \simeq -7.41$ m/s $\simeq -26.7$ km/h.

(2) The direct common sense approach is area conservation. The 13.7 m-high wall of water is advancing at a velocity close to $\sqrt{g\ell_c H} \simeq 18.6$ m/s $\simeq 66$ km/h, and presumably its forward motion must be fed by the undertow. Let $\ell \simeq 34.5$ m be the local undisturbed depth, $a \simeq 13.7$ m the bore height, and u the undertow velocity. Then area conservation (a bit naively applied) says that $-u(\ell - a) = \sqrt{g\ell}a$ and so $u = -\frac{1}{\ell - a}\sqrt{g\ell}$. To be consistent with the formal asymptotics of Problem 8.9, we evoke $\ell \ll a$; so $u \simeq -\frac{a}{\ell}\sqrt{g\ell} \simeq -26.2$ km/h, close to the estimate in (1). But if we insist on taking the $\ell - a$ seriously, we'd get a higher speed, $u \simeq -43.4$ km/h.

This example clearly "pushes the envelope" of simple asymptotic modeling. This would not be the first time, or the last, that a simple model is applied outside its domain of quantitative validity in order to glimpse qualitatively "what is out there".

Guide to bibliography. Recommended references for this chapter are Courant & Friedrichs [7], Lighthill [18], Ockendon & Ockendon [20], Stoker [22], and Whitham [24].

Lighthill [18], Ockendon & Ockendon [20], Stoker [22], and Whitham [24] all discuss the standard topics: simple waves, Riemann invariants, and characteristics in the context of the shallow water equations, as well as the jump condition across a bore, including the cubic dependence of energy dissipation on bore height. Lord Rayleigh is mentioned in connection with bore-induced energy dissipation, following Whitham [24].

One theme of this chapter is how Riemann invariants along characteristics crossing a weak shock undergo a change that scales like the shock strength cubed. This means that small-amplitude simple waves that break can still be treated asymptotically as simple waves after breaking. Lighthill [18] discusses this exact point qualitatively, and hence anticipates the formal asymptotic reduction to Burger's equation carried out in this chapter. Whitham [24] starts his book with Burger's equation as a pedagogical model, long before it is linked quantitatively to gas dynamics or shallow water theory. Courant & Friedrichs' book [7] is included in the reference list because it briefly mentions shock waves in elastic solids, touched upon in Problem 8.7.

Bibliography

[1] G.I. Barenblatt (1996) *Scaling, Self-Similarity, and Intermediate Asymptotics*, Cambridge University Press, Cambridge, UK.

[2] G.K. Batchelor (1967) *An Introduction to Fluid Mechanics*, Cambridge University Press, Cambridge, UK.

[3] Y.B. Chernyak, R.M. Rose (1995) *The Chicken from Minsk and 99 Other Infuriatingly Challenging Brainteasers from the Great Russian Tradition of Math and Science*, Basic Books, New York, NY.

[4] A.J. Chorin, J.E. Mardsen (1988) *A Mathematical Introduction to Fluid Mechanics*, Springer, New York, NY.

[5] A.J. Chorin, O.H. Hald (2006) *Stochastic Tools in Mathematics and Science*, Springer, New York, NY.

[6] R.V. Churchill, J.W. Brown (1987) *Fourier Series and Boundary Value Problems*, McGraw-Hill, New York, NY.

[7] R. Courant, K.O. Friedrichs (1999) *Supersonic Flow and Shock Waves*, Springer, New York, NY.

[8] R. Courant, D. Hilbert (1953) *Methods of Mathematical Physics, Volume I*, John Wiley & Sons, New York, NY.

[9] R. Courant, D. Hilbert (1989) *Methods of Mathematical Physics, Volume II*, John Wiley & Sons, New York, NY.

[10] A. Erdélyi (1962) *Operational Calculus and Generalized Functions*, Holt, Rhinehart and Winston, New York, NY.

[11] L.C. Evans (1998) *Partial Differential Equations*, American Mathematical Society, Providence, RI.

[12] R.P. Feynman, R.B. Leighton, M. Sands (1963) *The Feynman Lectures on Physics, Volume I*, Addison-Wesley, Reading, MA.

[13] R.P. Feynman, R.B. Leighton, M. Sands (1963) *The Feynman Lectures on Physics, Volume II*, Addison-Wesley, Reading, MA.

[14] I.M. Gel'fand, G.E. Shilov (1964) *Generalized Functions, Volume I: Properties and Operations*, Academic Press, New York, NY and London, UK.

[15] F. John (1982) *Partial Differential Equations*, Springer, New York, NY.

[16] L.D. Landau, E.M. Lifshitz (1976) *Mechanics. Course of Theoretical Physics, Volume 1*, 3rd edition, Butterworth-Heinemann, Oxford, UK.

[17] L.D. Landau, E.M. Lifshitz (1959) *Fluid Mechanics. Course of Theoretical Physics, Volume 6*, Pergamon Press, Oxford, UK.

[18] J. Lighthill (1978) *Waves in Fluids*, Cambridge University Press, Cambridge, UK.

[19] M.J. Lighthill (1958) *Introduction to Fourier Analysis and Generalised Functions*, Cambridge University Press, Cambridge, UK.

[20] H. Ockendon, J.R. Ockendon (2004) *Waves and Compressible Flow*, Springer, New York, NY.

[21] P.G. Saffman (1992) *Vortex Dynamics*, Cambridge University Press, Cambridge, UK.

[22] J.J. Stoker (1957) *Water Waves*, Interscience Publishers, New York, NY.

[23] N. Wax, ed. (1954) *Selected Papers on Noise and Stochastic Processes*, Dover, New York, NY.

[24] G.B. Whitham (1974) *Linear and Nonlinear Waves*, John Wiley & Sons, New York, NY.

Index

Titles in This Series

TITLES IN THIS SERIES

For a complete list of titles in this series, visit the
AMS Bookstore at **www.ams.org/bookstore/**.